Seismic Analysis of Structures and Equipment

Praveen K. Malhotra

Seismic Analysis of
Structures and Equipment

 Springer

Praveen K. Malhotra
StrongMotions Inc.
Sharon, MA, USA

ISBN 978-3-030-57860-2   ISBN 978-3-030-57858-9   (eBook)
https://doi.org/10.1007/978-3-030-57858-9

This book is dedicated to my professors who taught me to think intuitively and express clearly.

Preface

Every book needs a purpose. This book fills the gap between earthquake science and engineering by smoothly transitioning from ground motion prediction to structural analysis and design. This book also fills the gap between prescriptive and complex analyses. Prescriptive analyses rely on empirical rules which are constantly debated and routinely updated after significant earthquakes. Complex analyses rely on black box computer programs whose limitations are not understood by most engineers. This book provides a middle approach, guided by analyses, that are sufficiently accurate, transparent, and free of numerical errors. The goal of this book is to empower engineers to innovate and to confidently stand behind the results of their analyses.

A secondary purpose of this book is to help non-engineers understand the concepts of a good seismic design. Ideas are presented in a step-by-step intuitive manner to enable understanding. The only requirements for following this book are some knowledge of Engineering Mechanics and an open mind.

Chapter 1 discusses the characteristics of seismic loads which distinguish them from other loads such as gravity and wind. Chapter 2 discusses the uncertain nature of seismic loads and how design loads are derived for different types of structures. Chapters 3–9 present carefully selected examples which illustrate important concepts in seismic design. Preferably, these chapters should be read sequentially, but they can also be read independently of one another.

It is emphasized throughout this book that the seismic performance of a structure depends on its mass, strength, deformability, and damping. An engineer may not have much control over the mass of a structure, but an engineer can control a structure's strength, deformability, and damping. Reliance on strength alone, to improve seismic performance, can be too expensive or impractical. Indiscriminate increase in strength can sometimes reduce the deformability and damping of a structure, thereby resulting in a less-safe design. Deformability and damping cannot be taken for granted. They are ensured through proper detailing and calculated through proper analysis.

This book clarifies many misconceptions and introduces some new concepts such as energy demand, seismic toughness, and the toppling response spectrum.

Sharon, MA, USA   Praveen K. Malhotra

Acknowledgments

This book is based on a series of lectures presented in the USA and abroad on behalf of the American Society of Civil Engineers (ASCE).

The ideas in this book were consolidated after working on challenging projects for clients such as Ameren Corporation, American Tank & Vessels, Amgen Thousand Oaks, Amgen Manufacturing Limited, Atlas, Boeing, Dubai Municipality, Federal Energy Regulatory Commission (FERC), Haisla Nation, International Thermonuclear Experimental Reactor (ITER), Kinder Morgan, Trans Mountain Canada, and many others.

Professor Manish Kumar of IIT Gandhinagar carefully read the entire manuscript and provided helpful suggestions.

My wife Romee provided constant encouragement and support to complete this book.

Contents

Chapter 1
Ground Motions from Past Earthquakes

Nomenclature

ζ	Viscous damping ratio
ADRS	Acceleration–deformation response spectrum
ED	Energy demand (area within the ADRS)
EOM	Equation(s) of motion
g	$9.81\ m/s^2$ = acceleration due to gravity
k	Stiffness of SDOF system
m	Mass of SDOF system
M	Magnitude of earthquake
NRS	Normalized response spectrum
PD	Peak deformation
PF	Peak force
PGA	Peak ground acceleration
PGD	Peak ground displacement
PGV	Peak ground velocity
PGV_n	Normalized peak ground velocity
PPA	Peak pseudo-acceleration
PPV	Peak pseudo-velocity
PSE	Peak strain energy
SDOF	Single-degree-of-freedom
T	Natural period of structure
T_c	Central period of ground motion

© Springer Nature Switzerland AG 2021
P. K. Malhotra, *Seismic Analysis of Structures and Equipment*,
https://doi.org/10.1007/978-3-030-57858-9_1

1.1   Introduction

An earthquake is a rupture of the brittle rock in the outer skin (crust) of the earth [1]. The rupture is caused by the slow deformation of the rock due to the movement of lava (magma) below the crust. The rate of deformation determines the rate of earthquakes; it varies from place to place. The size (area) of the rupture can range from a few square meters to thousands of square kilometers. The strain energy released by the rupture determines the *magnitude* **M** of the earthquake; it correlates well with the rupture area. The energy is carried by seismic waves traveling in all directions [1]. When waves arrive at a location, the ground starts to shake. Engineers are interested in specific characteristics of ground motions; those are discussed in this chapter.

1.2   Amplitude of Ground Motion

Ground motions that are strong enough to damage structures or *liquefy* soils are known as *strong motions*. Strong motions are measured by accelerometers (acceleration sensors) in three perpendicular directions—usually east, up, and north [2]. Recorded accelerations are digitally filtered (corrected) to remove perceived "noise" [2].

Figure 1.1 shows the corrected acceleration histories of ground motion at a site in Castaic, California during the 1994 magnitude **M** 6.7 Northridge Earthquake [3]. The accelerations are expressed in units of g = 9.81 m/s^2 = acceleration due

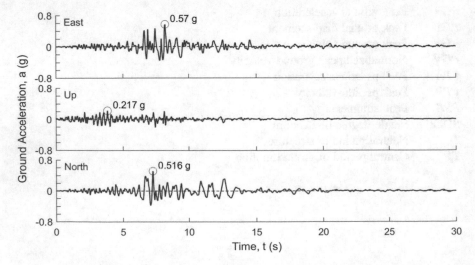

Fig. 1.1 Processed (corrected) acceleration histories at a site in Castaic, CA during the 1994 magnitude **M** 6.7 Northridge Earthquake in Southern California [3]

to gravity. The horizontal motion of the ground is of greater interest to engineers than the vertical motion because most structures can resist vertical (gravity) loads with high margin of safety. Apparent change in the gravity load due to vertical ground motion does not pose a problem for most structures, but there can be exceptions.

In Fig. 1.1, the maximum acceleration in the east direction is 0.570 g and the maximum acceleration in the north direction is 0.516 g. It is conceivable that the peak horizontal acceleration will not be in the east or the north directions, but some other direction in the horizontal plane. To determine the peak horizontal ground acceleration, the resultant accelerations are calculated from the accelerations in east and north directions. The instantaneous value of resultant horizontal acceleration is:

$$a_r = \sqrt{a_e^2 + a_n^2}$$

where a_e and a_n are instantaneous values of accelerations in east and north directions, respectively. The instantaneous direction of resultant acceleration (clockwise from north) is:

$$\theta = \tan^{-1} a_e / a_n$$

Figure 1.2 shows a plot of the resultant horizontal ground accelerations during the earthquake. The peak horizontal ground acceleration is $PGA = 0.574$ g.

The ground accelerations in Fig. 1.1 are numerically integrated to obtain ground velocities [2]. Numerical integration is easily performed by using software tools such as MATLAB [4]. Any "unreal" trends in the velocity histories are removed based on judgment. Figure 1.3 shows corrected velocity histories in east, up, and north directions. The resultant horizontal velocities are calculated in a similar manner as the resultant horizontal accelerations. Figure 1.4 shows resultant horizontal velocities during the earthquake. The peak horizontal ground velocity is $PGV = 58$ cm/s. The ground velocities in Fig. 1.3 are numerically integrated to obtain ground displacements [2]. Any "unreal" trends in the displacement histories are removed based on judgment. Figure 1.5 shows corrected displacement histories in east, up, and north directions. Figure 1.6 shows resultant horizontal displacements during the earthquake. The peak horizontal ground displacement is $PGD = 17.2$ cm.

Ground displacements derived from acceleration measurements may not be very accurate [2] because "slowly changing" displacements of the ground do not produce enough accelerations to be picked up by the accelerometers. Fortunately, slowly changing displacements of the ground do not induce significant response in most structures. Therefore, displacements derived from acceleration measurements are reasonable for most engineering analyses.

The amplitude of a ground motion is defined by three parameters—PGA, PGV, and PGD [5]. Sometimes, these are called *intensity measures* IM. A ground motion is composed of many different frequencies. It will become more clear later in this chapter that PGA, PGV, and PGD correlate with high-, medium-, and low frequencies in the ground motion. High value of PGA (Fig. 1.2) implies that the ground

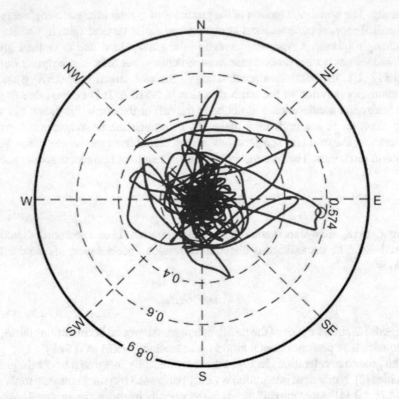

Fig. 1.2 Resultant horizontal accelerations

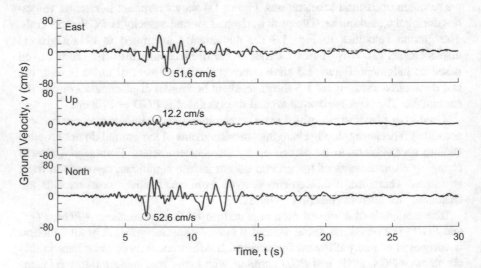

Fig. 1.3 Velocity histories in three orthogonal directions [3]

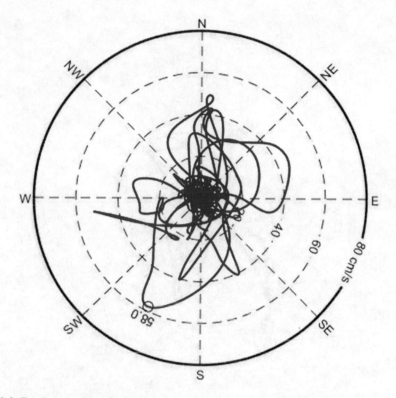

Fig. 1.4 Resultant horizontal velocities

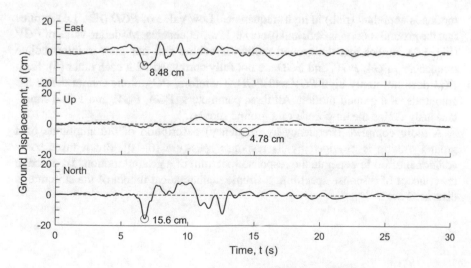

Fig. 1.5 Displacement histories in three orthogonal directions [3]

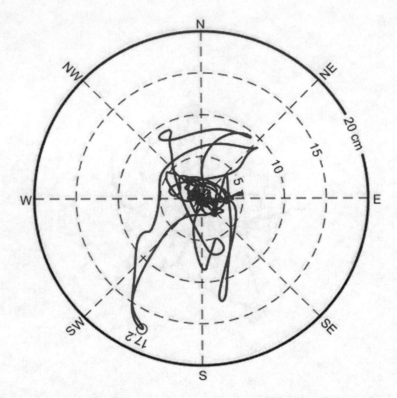

Fig. 1.6 Resultant horizontal displacements

motion is abundant (rich) in high frequencies. Low value of *PGD* (Fig. 1.6) implies that the ground motion is deficient (poor) in low frequencies. Moderate value of *PGV* (Fig. 1.4) implies that the ground motion is neither rich nor poor in intermediate frequencies. *PGA*, *PGV*, and *PGD* are not fully correlated with each other [5]. High *PGA* does not imply high *PGV* and *PGD*. Therefore, *PGA* alone cannot define the amplitude of a ground motion. All three parameters (*PGA*, *PGV*, and *PGD*) simultaneously define the amplitude of a ground motion.

A more complete (frequency-by-frequency) description of the amplitude of a ground motion is provided by its response spectrum [6–10]. But it takes some additional effort to generate the response spectrum of a ground motion. To describe the concept of response spectrum, a simple mathematical model of the structure is discussed next.

1.3   Single-Degree-of-Freedom System

Any real structure can be approximately represented by one or more single-degree-of-freedom (SDOF) systems of the type shown in Fig. 1.7 [6–10]. The deformed state of a SDOF system is completely defined by a single variable—spring deformation u. If the spring is linear with stiffness k, the force required to deform the spring by u_0 is ku_0. When the force is suddenly released, the mass m starts oscillating as shown in Fig. 1.8. The time taken to complete each oscillation cycle, T, is known as the natural period of the SDOF system. An increase in mass lengthens the natural period, and an increase in stiffness shortens the natural period. For a linear system,

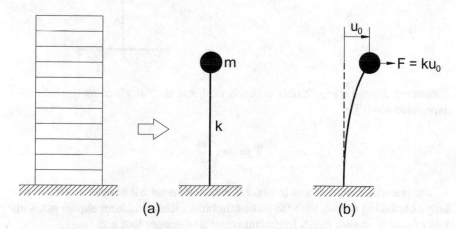

Fig. 1.7   SDOF model of a structure (e.g., a multistory building)

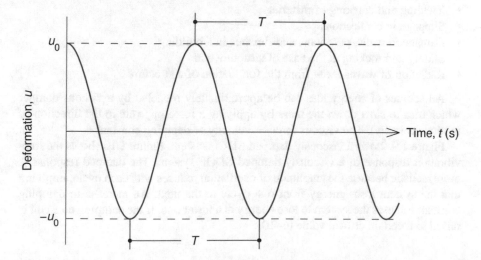

Fig. 1.8   Free-vibration response of SDOF system shown in Fig. 1.7

Fig. 1.9 Viscously damped
SDOF model of a structure

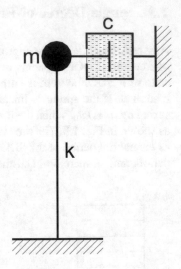

T does not depend on u_0, it only depends on k and m. T is given by the following expression [6–10]:

$$T = 2\pi\sqrt{\frac{m}{k}} \tag{1.1}$$

The free-vibration response in Fig. 1.8 is unreal because it implies that the system keeps oscillating forever with the same amplitude. Real structures experience some loss of energy in each cycle. Important sources of energy loss are:

- Air resistance
- Yielding and cracking of material
- Slippage in connections
- Damage to walls, partitions, and finishes in a building
- Sliding and rocking at the base of structure and
- Radiation of waves away from the foundation of a structure

All sources of energy loss can be approximately modeled by a viscous damper which tries to slow down the mass by applying a force opposite to the direction of motion (velocity). The viscous force = velocity × damping constant c.

Figure 1.9 shows a viscously damped SDOF system. Figure 1.10 shows the free-vibration response of a viscously damped SDOF system. The damped response is more realistic because the amplitude of oscillation reduces with each cycle, implying that the system loses energy from one cycle to the next. An increase in damping constant c causes the system to lose energy at a faster rate. If the damping constant is raised to a certain critical value [6–10]:

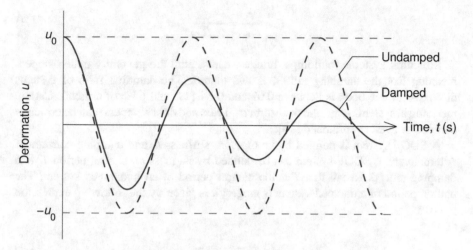

Fig. 1.10 Free-vibration responses of damped and undamped SDOF systems [10]

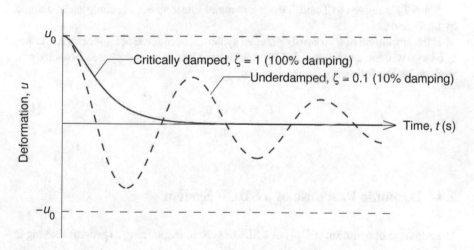

Fig. 1.11 Free-vibration responses of underdamped and critically damped SDOF systems

$$c_{\text{crit}} = 2\sqrt{km} \tag{1.2}$$

the system does not oscillate when released from the deformed position [6–10]. Figure 1.11 shows the free-vibration response of a critically damped system; the system arrives at the zero (undeformed) position from one side only—without crossing over to the other side. It is customary to express damping as a fraction (ratio) of the critical damping. The damping ratio is defined as:

$$\zeta = \frac{c}{c_{\text{crit}}} = \frac{c}{2\sqrt{km}} \tag{1.3}$$

Structural systems (buildings, bridges, dams, etc.) are generally underdamped; meaning that the damping ratio ζ is less than 1. The damping ratio of systems discussed in this book is between 0.02 and 0.64 (2% and 64% of critical). Certain mechanical systems (e.g., shock absorbers in cars and door stoppers) can be critically damped or even overdamped [10].

A SDOF system is defined by its mass m, stiffness k, and damping constant c. Alternatively, a SDOF system can be defined by its mass m, natural period T, and damping ratio ζ. Recall that T is the natural period of an undamped system. The natural period of a damped system is longer; it is given by the following expression [6–10]:

$$T_d = \frac{T}{\sqrt{1 - \zeta^2}} \tag{1.4}$$

Since T_d is related to T and ζ, even a damped linear system is completely defined by m, T, and ζ.

If the amplitude of a viscously damped system is kept fixed at u_0, a certain amount of energy will have to be supplied to replenish the loss. The energy loss in each cycle is given by the following expression:

$$E_l = 2\pi^2 \frac{c}{T} u_0^2 \tag{1.5}$$

1.4 Dynamic Response of a SDOF System

The equation of a motion (EOM) of a SDOF system responding to ground shaking is [7–10]:

$$m\ddot{u}(t) + c\dot{u}(t) + ku(t) = -ma(t) \tag{1.6}$$

where $u(t) =$ deformation of the system (spring) at time t, and $a(t) =$ acceleration of the ground at time t. The overdot represents differentiation with respect to time. Dividing throughout by m and making use of Eqs. 1.1 and 1.3, the EOM of a SDOF system can be rewritten as:

$$\ddot{u}(t) + \frac{4\pi\zeta}{T}\dot{u}(t) + \left(\frac{2\pi}{T}\right)^2 u(t) = -a(t) \tag{1.7}$$

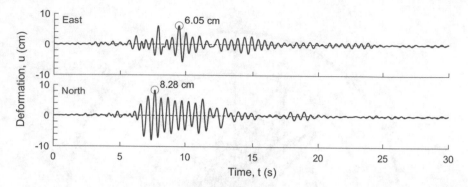

Fig. 1.12 Deformations of a SDOF system of period $T = 0.5$ s, and damping $\zeta = 0.05$ (5% of critical), subjected to horizontal ground motions shown in Fig. 1.1

It is apparent from Eq. 1.7 that for a given acceleration history $a(t)$, the deformation history $u(t)$ depends only on the natural period T and the damping ratio ζ. Equation 1.7 is a second-order linear differential equation which can be numerically solved by using MATLAB [4].

Consider a SDOF system of period $T = 0.5$ s and damping $\zeta = 0.05$ (5% of critical). The system is subjected to the east-direction ground acceleration history shown in Fig. 1.1. Equation 1.7 is solved for deformation history $u(t)$ using MATLAB [4]. The upper part of Fig. 1.12 displays the east-direction deformation history. Next, the same SDOF system is subjected to the north-direction acceleration history shown in Fig. 1.1. The north-direction deformation history is displayed in the lower part of Fig. 1.12. At any given time, the SDOF system has deformed in both east and north directions. The amplitude and direction of resultant deformation are computed at various times from the instantaneous deformations in east and north directions. Figure 1.13 shows a plot of the resultant horizontal deformations of the SDOF system during the earthquake. The peak value of resultant horizontal deformation of 0.5 s period, 5% damped system is $PD(0.5$ s$) = 9.38$ cm.

Next, the period of the SDOF system is increased to $T = 1$ s, but the damping is maintained at $\zeta = 0.05$ (5% of critical). Figure 1.14 shows the resultant horizontal deformations during the earthquake. The peak resultant horizontal deformation of the 1 s period, 5% damped system is $PD(1$ s$) = 24.7$ cm. Next, the period of the system is changed to $T = 2$ s, but the damping is still maintained at $\zeta = 0.05$ (5% of critical). Figure 1.15 shows the resultant horizontal deformations during the earthquake. The peak resultant horizontal deformation of the 2 s period, 5% damped system is $PD(2$ s$) = 33.4$ cm.

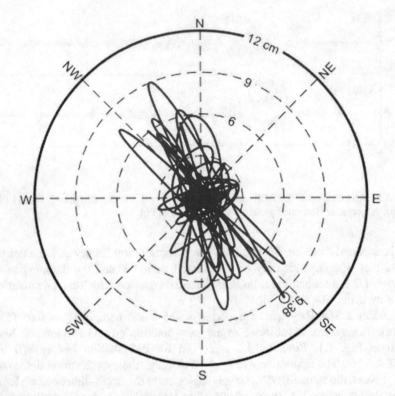

Fig. 1.13 Resultant horizontal deformations of SDOF system of 0.5 s period and 5% damping, subjected to the horizontal ground motions shown in Fig. 1.1

1.5 Deformation Response Spectrum

The deformation response spectrum of a ground motion is a plot between the natural period T and peak deformation PD, for a fixed value of damping ζ. Figure 1.16 shows the 5% damping deformation response spectrum of ground motion shown in Fig. 1.1. Note that $PD(T = 0) = 0$. According to Eq. 1.1, a zero-period system is either infinitely stiff or it has no mass at all; such a system is not expected to deform during ground shaking. With increase in T, the peak deformation PD generally increases at first and then decreases to approach the peak ground displacement PGD at $T = \infty$. According to Eq. 1.1, a system with infinite period has infinite mass or zero stiffness. For such a system, the mass remains stationary while the ground moves underneath, or the deformation of the spring equals the ground displacement. Therefore, for an infinite-period system, $PD = PGD$, irrespective of damping.

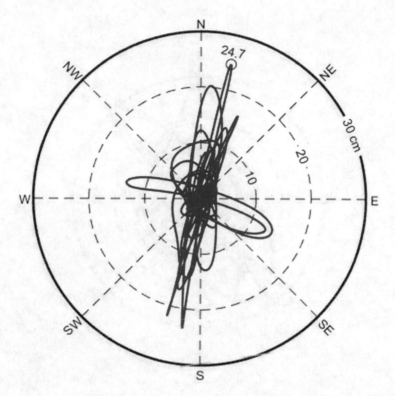

Fig. 1.14 Resultant horizontal deformations of SDOF system of 1 s period and 5% damping, subjected to the horizontal ground motions shown in Fig. 1.1

1.6 Pseudo-Acceleration Response Spectrum

The peak force in the spring is the product of peak deformation and spring stiffness:

$$PF = k \cdot PD \tag{1.8}$$

With the help of Eq. 1.1, the stiffness k can be expressed in terms of m and T to yield the following expression for the peak force:

$$PF = m \cdot \left(\frac{2\pi}{T}\right)^2 \cdot PD \tag{1.9}$$

The multiplier of mass m in the above expression has the units of acceleration, but it is only approximately equal to the "true" peak acceleration of the mass. Therefore, it is simply called the peak pseudo-acceleration *PPA*, defined as follows:

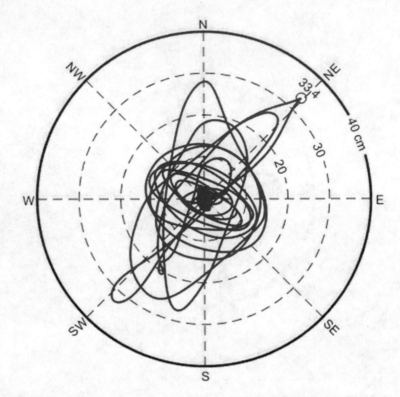

Fig. 1.15 Resultant horizontal deformations of SDOF system of 2 s period and 5% damping, subjected to the horizontal ground motions shown in Fig. 1.1

$$PPA = \left(\frac{2\pi}{T}\right)^2 \cdot PD \qquad (1.10)$$

The expression for the peak force in the spring (Eq. 1.9) can be rewritten as:

$$PF = m \cdot PPA \qquad (1.11)$$

Note that the "true" force in the spring is obtained by multiplying the mass by the "pseudo-acceleration."

A plot of *PPA* versus *T* is known as the pseudo-acceleration response spectrum. A pseudo-acceleration response spectrum can be generated from a deformation response spectrum simply by using Eq. 1.10. For example, $PD(1\ s) = 24.7$ cm. Therefore, $PPA(1\ s) = (2\pi/1)^2 \times 24.7 = 974\ \text{cm/s}^2 = 0.993$ g. *PPA* values can be similarly calculated for other periods. Figure 1.17 shows a plot of the 5% damping pseudo-acceleration response spectrum of ground motion shown in Fig. 1.1. It is customary to express pseudo-accelerations in units of g (acceleration due to gravity). Note that $PPA(T = 0) = PGA$. A zero-period system is so stiff that it does not deform; it moves in unison with the ground. Because the system does not deform, the

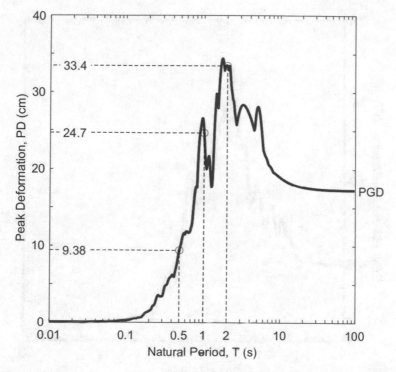

Fig. 1.16 Five-percent damping deformation response spectrum of horizontal ground motion shown in Fig. 1.1

damping force is zero and the peak force in the spring is $m \cdot PGA$, or $PPA = PGA$. For this reason, PGA is sometimes known as zero-period acceleration ZPA. For an infinite-period system, the spring has zero stiffness, therefore the spring force is zero or $PPA(T = \infty) = 0$.

1.7 Pseudo-Velocity Response Spectrum

The peak strain energy in the spring is:

$$PSE = \frac{1}{2}k \cdot PD^2 \tag{1.12}$$

With the help of Eq. 1.1, the stiffness k can be expressed in terms of m and T to yield the following expression for the peak strain energy:

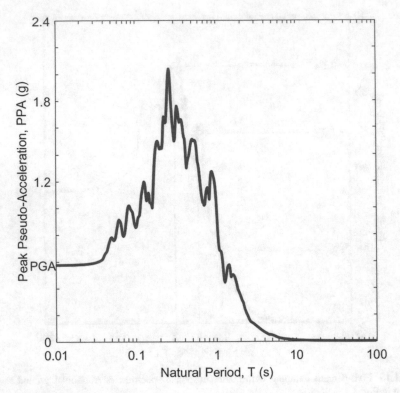

Fig. 1.17 Five-percent damping pseudo-acceleration response spectrum of horizontal ground motion shown in Fig. 1.1

$$PSE = \frac{1}{2}m \cdot \left(\frac{2\pi}{T} \cdot PD\right)^2 \tag{1.13}$$

The term in parentheses in the above expression has the units of velocity, but it is not the true velocity of the mass [7, 8, 10]. Therefore, it is simply called the peak pseudo-velocity PPV, defined as:

$$PPV = \frac{2\pi}{T} \cdot PD \tag{1.14}$$

The expression for the peak strain energy (Eq. 1.13) can be rewritten as:

$$PSE = \frac{1}{2}m \cdot PPV^2 \tag{1.15}$$

Note that the "true" strain energy in the spring is obtained by multiplying half the mass by the square of the "pseudo-velocity."

A plot of PPV versus T is known as the pseudo-velocity response spectrum. A pseudo-velocity response spectrum can be generated from a deformation response

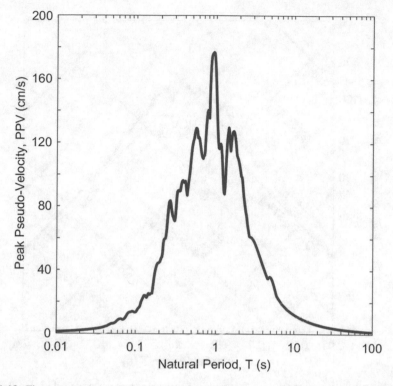

Fig. 1.18 Five-percent damping pseudo-velocity response spectrum of horizontal ground motion shown in Fig. 1.1

spectrum simply by using Eq. 1.14. For example, $PD(1\ s) = 24.7$ cm. Therefore, $PPV(1\ s) = (2\pi/1) \times 24.7 = 155$ cm/s. PPV values can be similarly calculated for other periods. Figure 1.18 shows a 5% damping pseudo-velocity response spectrum of ground motion shown in Fig. 1.1. Note that $PPV = 0$ for $T = 0$ as well as for $T = \infty$. A zero-period system is so stiff that it does not deform hence does not store any strain energy. An infinite-period system is so flexible that its spring lacks any stiffness to store strain energy.

1.8   Tripartite Response Spectrum

PD, PPA, and PPV are related to each other as follows:

$$PPA \cdot \frac{T}{2\pi} = PPV = \frac{2\pi}{T} \cdot PD \tag{1.16}$$

Therefore, it is possible to display all three response spectra (deformation, pseudo-acceleration, and pseudo-velocity) by a single curve in a tripartite format.

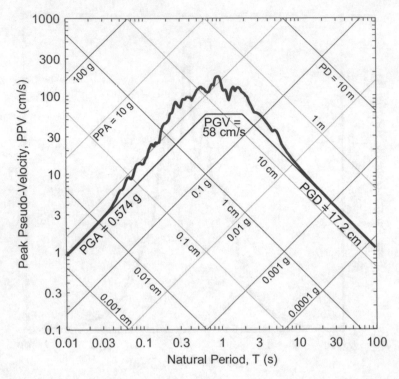

Fig. 1.19 Tripartite plot of 5% damping response spectrum of horizontal ground motion shown in Fig. 1.1

Figure 1.19 displays the 5% damping tripartite response spectrum of horizontal ground motion shown in Fig. 1.1. In this plot, *PPV* is read along the vertical axis and period *T* is read along the horizontal axis. The peak pseudo-acceleration *PPA* is read along the −45° (counterclockwise) axis, and peak deformation *PD* is read along the +45° (clockwise) axis. For reference, *PGA*, *PGV*, and *PGD* are shown by three straight lines. Tripartite plot clearly shows that *PPA* approaches *PGA* at very short periods and *PD* approaches *PGD* at very long periods. The tripartite plot is a very concise and elegant way of displaying the response spectrum of a ground motion.

1.9 Central Period and Normalized Velocity

Ground motions are composed of many different frequencies. The central period of a ground motion is defined as follows [11]:

$$T_c = 2\pi \sqrt{\frac{PGD}{PGA}} \tag{1.17}$$

The central period is not the dominant period of a ground motion; it is simply the "centroid" of all frequencies present in the ground motion. If the ground motion was composed of a single frequency, that frequency would be $2\pi/T_c$. In Fig. 1.19, if PGA and PGD lines were extended, they will intersect at period T_c. Central period T_c can be used to classify a ground motion as low-, medium-, or high frequency, as per Table 1.1.

If a ground motion was composed of a single frequency, its peak velocity PGV will be equal to $\sqrt{PGA \cdot PGD}$. Since a ground motion is usually composed of many different frequencies, the actual peak velocity is less than $\sqrt{PGA \cdot PGD}$. The normalized-velocity PGV_n is the ratio between the actual peak ground velocity PGV and $\sqrt{PGA \cdot PGD}$, i.e.,

$$PGV_n = \frac{PGV}{\sqrt{PGA \cdot PGD}} \tag{1.18}$$

Normalized-velocity PGV_n is an indicator of the frequency band of a ground motion [11]. High value of PGV_n implies a narrow-banded ground motion, and a low value of PGV_n implies a broad-banded ground motion. Table 1.2 classifies ground motions based on PGV_n.

It will become clear in Sect. 2.10 that the strength of a structure is most effective in resisting low-frequency ground motions, deformability of a structure is most effective in resisting high-frequency ground motions, and damping of a structure is most effective in resisting narrow-banded ground motions. Therefore, T_c and PGV_n of future ground motions are helpful in selecting the type of structure for a site.

Table 1.1 Frequency classification of a ground motion based on its central period T_c

T_c	Frequency classification
<0.5 s	High frequency
0.5 s to 2 s	Medium frequency
>2 s	Low frequency

Table 1.2 Frequency band of a ground motion based on its normalized-velocity PGV_n

PGV_n	Frequency band
<0.45	Broad
0.45–0.75	Medium
>0.75 s	Narrow

1.10 Response Spectrum of Incompatible Acceleration, Velocity, and Displacement Histories

Processed histories of ground acceleration, velocity, and displacement are sometimes not fully compatible with each other [12]. Meaning that the integration and double integration of the acceleration history do not produce the processed velocity and displacement histories. In such cases, the response spectrum is generated from the acceleration history at short periods and velocity and displacement histories at long periods. The response spectrum generated in this manner shows the correct asymptotic behavior at both short and long periods [12].

1.11 Smooth Response Spectrum of Ground Motion

The response spectra discussed so far are generated from ground motion histories. However, for future earthquakes, ground motion histories cannot be directly predicted. Only parameters such as PGA, PGV, and PGD can be predicted, as will be seen in the next chapter. Therefore, it is desirable to generate the response spectrum from predicted values of PGA, PGV, and PGD.

Spectral values (PPA, PPV, and PD) for short periods are controlled by high frequencies in ground motion. High frequencies also determine the value of PGA. Therefore, short-period spectral values correlate with PGA [5, 11]. Similarly, long-period spectral values correlate with PGD and intermediate-period spectral values correlate with PGV [5, 11]. Therefore, it is possible to construct a smooth response spectrum from predicted values of PGA, PGV, and PGD.

The response spectrum relative to PGA, PGV, and PGD is known as the normalized response spectrum (NRS) [11]. The shape of the NRS was established using thousands of records from past earthquakes. It was found that the shape of the NRS depends only on the normalized velocity PGV_n, it does not depend on any other parameters such as the magnitude of the earthquake, distance of the earthquake, or the local soil conditions. Most importantly, the shape of the NRS does not depend on the direction of ground motion; it is same for both horizontal and vertical ground motions. Figures 1.20 and 1.21 show NRS for $PGV_n = 0.3$ (broad-banded ground motion) and $PGV_n = 0.9$ (narrow-banded ground motion). In Figs. 1.20 and 1.21, the horizontal axis is the normalized period T/T_c and the vertical axis is the normalized peak pseudo-velocity $PPV/\sqrt{PGA \cdot PGD}$. In Table 1.3, the normalized periods T/T_c are listed in the first column and normalized peak pseudo-velocities $PPV/\sqrt{PGA \cdot PGD}$ are listed in columns 2–8 for various values of normalized-velocity PGV_n.

The following example illustrates the generation of a smooth response spectrum from given values of PGA, PGV, and PGD.

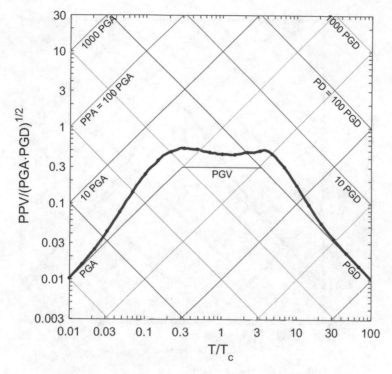

Fig. 1.20 Normalized response spectrum for $PGV_n = 0.3$ (broad-banded ground motion)

1.12   Example: Smooth Response Spectrum from *PGA*, *PGV*, and *PGD*

For a specific site, the predicted values of *PGA*, *PGV*, and *PGD* are 0.3 g (294 cm/s^2), 50 cm/s, and 30 cm, respectively. It is of interest to develop a 5% damping response spectrum for the site. The following steps are taken to generate the site response spectrum:

1. The central period of the ground motion is calculated from Eq. 1.17:

$$T_c = 2\pi\sqrt{PGD/PGA} = 2\pi\sqrt{30/294} = 2 \text{ s}.$$

According to Table 1.1, the ground motion is medium frequency.
2. The normalized velocity is calculated from Eq. 1.18:

$$PGV_n = PGV/\sqrt{PGA \times PGD} = 50/\sqrt{294 \times 30} = 0.53.$$

According to Table 1.2, the ground motion is medium banded.

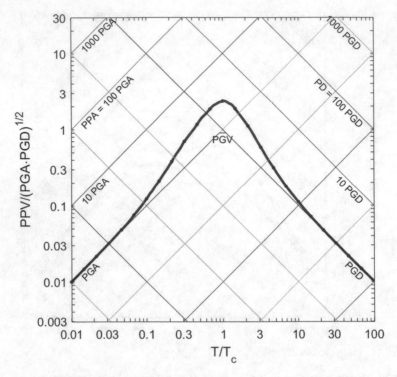

Fig. 1.21 Normalized response spectrum for $PGV_n = 0.9$ (narrow-banded ground motion)

3. The normalized periods are read from the first column of Table 1.3. These are multiplied by the central period $T_c = 2$ s to obtain the natural periods T.
4. The normalized pseudo-velocities $PPV/\sqrt{PGA \cdot PGD}$ are read from the fourth column of Table 1.3 corresponding to $PGV_n = 0.5$. These are multiplied by $\sqrt{PGA \cdot PGD} = 94$ cm/s to obtain the peak pseudo-velocities PPV.

Figure 1.22 shows a plot PPV versus T. This is the 5% damping smooth response spectrum generated from PGA, PGV, and PGD. Next, PPA and PD are determined from PPV and T by using Eq. 1.16. Figure 1.23 shows pseudo-acceleration and deformation response spectra for the site. Note that the peak in the pseudo-acceleration response spectrum occurs at $T = 0.55$ s, while the peak in the deformation response spectrum occurs at $T = 6$ s. None of these peaks occur at the central period of the ground motion. This is expected of a medium-banded ground motion. Only for a narrow-banded ground motion, the peaks in the pseudo-acceleration and the deformation response spectra occur at the central period of the ground motion. It is incorrect to call 0.55 s or 6 s as the "site period."

Table 1.3 Peak normalized pseudo-velocities $PPV/\sqrt{PGA \times PGD}$ for 5% damping [11]

$\frac{T}{T_c}$	PGV_n						
	0.3	0.4	0.5	0.6	0.7	0.8	0.9
0.01	0.01	0.01	0.01	0.01	0.01	0.01	0.01
0.0147	0.0157	0.0153	0.0149	0.0149	0.0147	0.0147	0.0147
0.0215	0.0247	0.0233	0.0221	0.0221	0.0218	0.0217	0.0216
0.0316	0.0428	0.0384	0.0346	0.0338	0.0332	0.0324	0.0322
0.0464	0.077	0.0669	0.0584	0.054	0.0525	0.0506	0.0483
0.0681	0.141	0.119	0.104	0.09	0.0889	0.0795	0.0753
0.1	0.245	0.214	0.182	0.16	0.154	0.13	0.123
0.147	0.369	0.351	0.325	0.289	0.277	0.235	0.212
0.215	0.478	0.542	0.532	0.506	0.486	0.426	0.381
0.316	0.532	0.691	0.782	0.821	0.8	0.759	0.7
0.464	0.517	0.753	0.962	1.1	1.22	1.21	1.17
0.681	0.472	0.724	1.01	1.32	1.48	1.77	1.85
1	0.449	0.68	0.974	1.34	1.56	2.11	2.36
1.47	0.443	0.668	0.975	1.32	1.57	1.8	1.87
2.15	0.475	0.705	0.898	1.1	1.15	1.15	1.11
3.16	0.496	0.661	0.752	0.752	0.726	0.675	0.583
4.64	0.452	0.47	0.479	0.426	0.387	0.361	0.314
6.81	0.304	0.283	0.252	0.227	0.208	0.195	0.179
10	0.175	0.156	0.137	0.125	0.119	0.115	0.112
14.7	0.0959	0.0894	0.0787	0.078	0.0729	0.0728	0.0712
21.5	0.0547	0.0531	0.0497	0.0494	0.0481	0.0477	0.0474
31.6	0.0341	0.0335	0.0326	0.0327	0.0323	0.0320	0.032
46.4	0.0224	0.0222	0.0219	0.0219	0.0217	0.0217	0.0216
68.1	0.015	0.0149	0.0148	0.0148	0.0147	0.0147	0.0147
100	0.01	0.01	0.01	0.01	0.01	0.01	0.01

1.13   Comparison with Newmark–Hall Response Spectrum

The idea of generating a smooth response spectrum from *PGA*, *PGV*, and *PGD* was first proposed by Newmark and Hall [8]. The Newmark–Hall spectrum was based on a limited number of records available up to the 1970s. The main shortcoming of the Newmark–Hall spectrum is that its shape does not depend on the normalized velocity. The Newmark–Hall spectrum has the same shape for both narrow-banded and board-banded ground motions. Figure 1.24 shows a comparison of the smooth spectrum generated in previous example with the Newmark–Hall response spectrum. In this case, the two spectra agree well because the ground motion is medium banded. For a broad-banded or a narrow-banded ground motion, the Newmark–Hall spectrum will be significantly different from the smooth spectrum presented in this chapter.

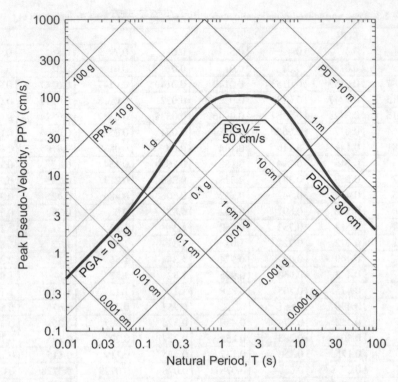

Fig. 1.22 Tripartite plot of 5% damping smooth response spectrum generated from *PGA*, *PGV*, and *PGD*

1.14   Building Code Response Spectra

The design response spectra in building codes [13] are based on the Newmark–Hall spectrum irrespective of the bandwidth of ground motion. Instead of *PGA* and *PGV*, code spectra utilize pseudo-accelerations at 0.2 s and 1 s period. Instead of *PGD*, code spectra utilize the long-period transition T_L. The current codes are beginning to recognize the limitation of the Newmark–Hall spectrum to capture the true shape of the response spectrum. For certain soil types, building codes [13] now require that a multi-period response spectrum be generated to capture the deviation from the Newmark–Hall spectrum. A multi-period response spectrum will require pseudo-accelerations at numerous periods. Therefore, the smooth spectrum discussed in this chapter is superior because it can be generated from just three parameters: *PGA*, *PGV*, and *PGD*. It is sometimes argued that *PGD* cannot be reliably predicted for future earthquakes, but the same argument can be applied to long-period transition T_L because *PGD* and T_L are related to each other.

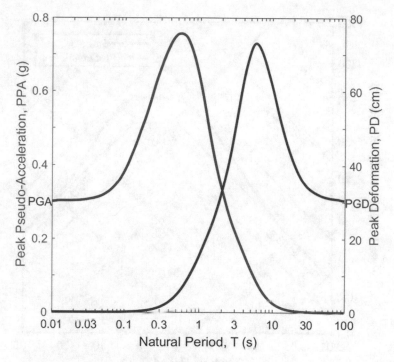

Fig. 1.23 Five-percent damping pseudo-acceleration and deformation response spectra generated from *PGA*, *PGV*, and *PGD*

1.15   Acceleration–Deformation Response Spectrum

For engineering applications, *PPA* and *PD* are much more useful than *PPV*. Therefore, it is desirable to show *PPA* and *PD* along the two principal axes (vertical and horizontal). This is achieved by rotating the tripartite response spectrum (Figure 1.22) clockwise by 45°. The resulting plot is shown in Figure 1.25; it has *PPA* along the vertical axis and *PD* along the horizontal axis. The period is indicated by parallel diagonal lines. Next, the scale on the plot is changed from logarithmic to linear, as shown in Figure 1.26. This causes the parallel period lines to meet at the origin. For the sake of clarity, the period lines are replaced by radial tick marks in Figure 1.26. The plots in Figures 1.25 and 1.26 are known as the acceleration–deformation response spectrum or simply ADRS. Such plots of the response spectra are very useful in "performance-based" seismic design. They will be extensively used in the remainder of this book.

Figure 1.27 illustrates how the response of a 1.5 s period system is read from the ADRS. A radial line is drawn from the origin to the tick mark corresponding to 1.5 s period. From the intersection of the radial line with the 5% damped ADRS, *PPA* = 0.44 g is read along the vertical axis and *PD* = 25 cm is read along the horizontal axis.

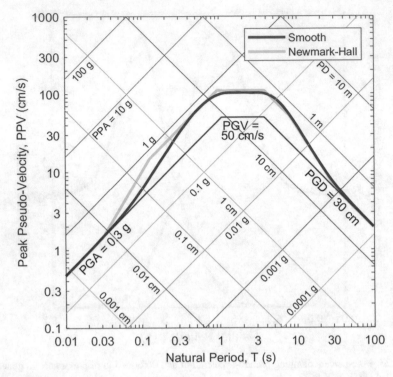

Fig. 1.24 Comparison of smooth response spectrum with the Newmark–Hall spectrum

1.16   Strength and Deformation Demands

The pseudo-acceleration in g's multiplied by the weight of the structure gives the lateral force induced in the structure. This is also the strength demand imposed by the ground motion. The other demand imposed by the ground motion is deformation. In Figure 1.28, the response spectrum of Figure 1.27 is redrawn by labeling the vertical and horizontal axes as "normalized strength" and "deformation demand." The plot in Figure 1.28 provides a useful insight into the seismic performance of structures. A structure with a lateral strength of 10% its weight needs to be able to deform 72 cm, while a structure with a lateral strength of 20% its weight needs to be able to deform only 54 cm. The strength and deformation demands are reciprocal to each other. A structure with high deformability need not be very strong, while a structure with low deformability needs to be very strong. The deformability is the ability of a structure to deform without getting collapsed. The most economical seismic design is achieved by relying partly on strength and partly on deformability. A purely strength-based seismic design is usually uneconomical. In high seismic regions, engineers usually trade strength for deformability.

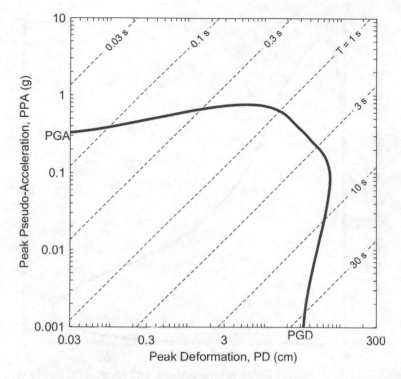

Fig. 1.25 Acceleration–deformation plot of response spectrum (ADRS) on a logarithmic scale

1.17 Unique Nature of Seismic Loads

Seismic loads are different from wind and gravity loads. Wind and gravity loads are monotonic; they continue to act in the same direction for a long time. Seismic loads are cyclic; they change direction very rapidly. Because of this difference, the resistance to wind and gravity loads can only be increased by increasing a structure's strength. The resistance to seismic load can be increased by increasing a structure's strength, deformability or damping. The role of damping will become clear in the next chapter.

1.18 Energy Demand *ED*

There is no single parameter which completely defines the seismic demand imposed by a ground motion, but the area enclosed within the 5% damping ADRS comes close to defining the seismic demand for a broad range of structures. For the response spectrum shown in Figure 1.28, $ED = 2.59$ (m/s)^2. *ED* has the units of energy/mass.

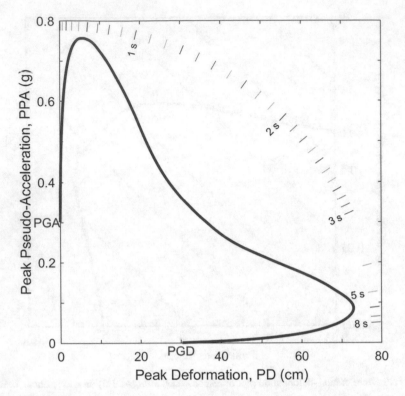

Fig. 1.26 Acceleration–deformation plot of response spectrum (ADRS) on a linear scale

1.19   Summary

1. The amplitude of a ground motion is not defined by *PGA* alone; it is defined by three parameters *PGA*, *PGV*, and *PGD*. *PGA* controls the response (deformation) of stiff structures, *PGD* controls the response of flexible structures, and *PGV* controls the response of structures that are neither stiff nor flexible.
2. The response spectrum provides a more complete description of the amplitude of ground motion than *PGA*, *PGV*, and *PGD*.
3. The exact response spectrum of a ground motion can be generated from the numerical solution of the EOM. A smooth response spectrum of ground motion can be generated from *PGA*, *PGV*, and *PGD*; it cannot be generated from *PGA* alone.
4. The response spectrum relative to *PGA*, *PGV*, and *PGD* is known as the normalized response spectrum (NRS). The shape of the NRS depends only on the normalized velocity $PGV_n = PGV/\sqrt{PGA \cdot PGD}$.
5. The histories of future ground motions cannot be predicted directly, but *PGA*, *PGV*, and *PGD* of future ground motions can be predicted, as will be seen in the

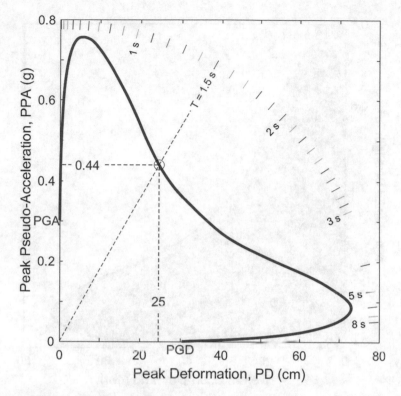

Fig. 1.27 Response of 1.5 s period SDOF system read from the ADRS

next chapter. Therefore, the NRS is useful in generating the response spectra of future ground motions.

6. An acceleration–deformation plot of the response spectrum clarifies the strength and deformation demands imposed by the ground motion.

7. Response spectrum is a property of the ground motion; it is not a property of the structure.

8. The strength and deformation demands are reciprocal to each other. Deformable structures need not be as strong as less deformable (brittle) structures.

9. Seismic loads are different from wind and gravity loads. The resistance to wind and gravity loads can only be increased by increasing a structure's strength. Seismic performance of a structure can be increased by increasing its strength, deformability or damping.

10. Strength of a structure is most effective in resisting low-frequency ground motions; and deformability of a structure is most effective in resisting high-frequency ground motions. Characteristics of future ground motions are helpful in selecting the type of structure for a site.

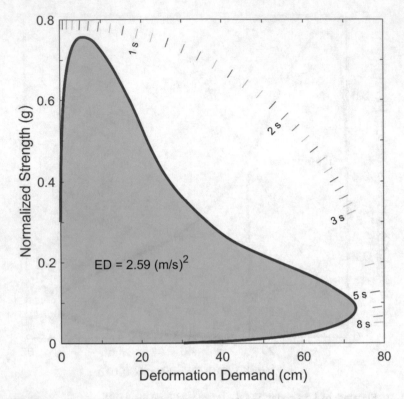

Fig. 1.28 Strength and deformation demands imposed by the ground motion

References

1. Bolt, B. (2006). *Earthquakes* (5th ed.). New York: W. H. Freeman and Company.
2. Hudson, D. E. (1979). *Reading and interpreting strong motion accelerograms*. Richmond, CA: EERI.
3. Darragh, R., Cao, T., Cramer, C., Graizer, V., Huang, M., & Shakal, A. (1994). *Processed CSMIP strong-motion data from the Northridge California Earthquake of 17 January 1994: Release No. 4*. Sacramento, CA: California Strong Motion Instrumentation Program (CSMIP).
4. MathWorks. (2020). *MATLAB Version 9.8.0.1417392 (R2020a)*. Natick, MA: MathWorks.
5. Malhotra, P. K. (2006). Smooth spectra of horizontal and vertical ground motions. *Bulletin of the Seismological Society of America, 96*(2), 506–518.
6. Biggs, J. M. (1964). *Introduction to structural dynamics*. McGraw Hill Book Company.
7. Clough, R. W., & Penzien, J. (1976). *Dynamics of structures*. McGraw Hill Publications.
8. Newmark, N. M., & Hall, W. (1982). *Earthquake spectra and design*. Richmond, CA: EERI.
9. Humar, J. L. (2012). *Dynamics of structures* (3rd ed.). CRC Press.
10. Chopra, A. K. (2017). *Dynamics of structures* (5th ed.). Pearson Publication.

11. Malhotra, P. K. (2015). Normalized response spectrum of ground motion. The Bridge & Structural Engineer. *Journal of the Indian National Group of the International Association for Bridge & Structural Engineering, 45*(1).
12. Malhotra, P. K. (2001). Response spectrum of incompatible acceleration, velocity and displacement histories. *Journal of Earthquake Engineering & Structural Dynamics, 30*(2), 279–286.
13. ASCE. (2016). *Minimum design loads for buildings and other structures* (ASCE/SEI 7–16). Reston, VA: American Society of Civil Engineers.

Chapter 2
Ground Motions for Future Earthquakes

Nomenclature

λ	Rate of exceedance
ζ	Viscous damping ratio
2D	Two dimensional
3D	Three dimensional
ADRS	Acceleration–deformation response spectrum
ED	Energy demand (area within the ADRS)
EOM	Equation(s) of motion
g	9.81 m/s^2 = acceleration due to gravity
GMPM	Ground motion prediction model
M	Moment magnitude of earthquake
MRI	Mean recurrence interval
MRP	Mean return period
PD	Peak deformation
PGA	Peak ground acceleration
PGA_V	Peak ground acceleration of vertical motion
PGD	Peak ground displacement
PGD_V	Peak ground displacement of vertical motion
PGV	Peak ground velocity
PGV_V	Peak ground velocity of vertical motion
PPA	Peak pseudo-acceleration
PPV	Peak pseudo-velocity
PSHA	Probabilistic seismic hazard analysis
SDOF	Single-degree-of-freedom system
SFBA	San Francisco Bay Area
T	Natural period of structure

© Springer Nature Switzerland AG 2021
P. K. Malhotra, *Seismic Analysis of Structures and Equipment*,
https://doi.org/10.1007/978-3-030-57858-9_2

2.1 Introduction

In Chap. 1, the response spectrum was shown to define the strength and deformation demands imposed by the ground motion. The response spectrum can be exactly computed from ground motion histories or it can be approximately generated from the peak values of ground acceleration, ground velocity, and ground displacement *PGA*, *PGV*, and *PGD*. Engineers need response spectra of ground motions produced by future earthquakes. For future earthquakes, ground motion histories cannot be directly predicted, but *PGA*, *PGV*, and *PGD* can be predicted although with high uncertainty. This chapter discusses the prediction of *PGA*, *PGV*, and *PGD* and generation of site-specific response spectra and ground motion histories for future earthquakes.

2.2 Prediction of *PGA*, *PGV*, and *PGD* at a Site

Three types of information are needed to predict *PGA*, *PGV*, and *PGD* at a site due to future earthquakes:

1. **Geological information** regarding the size (magnitude), location, and occurrence rate of future earthquakes in the region.
2. **Seismological information** regarding *PGA*, *PGV*, and *PGD* at a site due to earthquakes of known magnitude occurring at known distance from the site.
3. **Geotechnical information** regarding the type and depth of soil at the site.

The uncertainties in geological and seismological models are so high that the predictions of *PGA*, *PGV*, and *PGD* can only be made probabilistically. The analysis that combines the uncertainties in geological and seismological models is known as the probabilistic seismic hazard analysis (PSHA) [1–5]. Next, the PSHA of a site is illustrated with the help of an example.

2.2.1 Example: PSHA of a Site

The local soil conditions are approximately defined by the average shear-wave velocity in the top 30 m (~100 ft) V_{S30}. Geotechnical investigation of the site shows that $V_{S30} = 275$ m/s. The site is classified as "stiff soil" (Site Class D) [6] by the geotechnical engineer.

Geological investigation has identified two principal sources of earthquakes in the region. These are *fault* lines where the rock has broken in the past and is expected to break again in the future. Fault 1 is 50 km long and its minimum distance from the site is 15 km. Based on its length, Fault 1 is estimated to produce earthquakes of magnitude **M** 6. Based on the rate of strain accumulation around Fault 1, the occurrence rate of **M** 6 earthquakes is estimated to be 0.004/year or the mean recurrence interval is MRI = 1/0.004 = 250 years. Fault 2 is 250 km long and its minimum distance from the site is 60 km. Based on its length, Fault 2 is estimated to produce earthquakes of magnitude **M** 7.5. Based on the rate of strain accumulation around Fault 2, the occurrence rate of **M** 7.5 earthquakes is 0.002/year or the mean recurrence interval is MRI = 1/0.002 = 500 years. Table 2.1 summarizes the geological data. Figure 2.1 shows the seismic sources (faults) relative to the site.

Based on the seismological model (also known as the ground motion prediction model GMPM), the expected (mean) values of *PGA*, *PGV*, *PGD* at the site due to earthquakes on Fault 1 and Fault 2 are listed in column 3 of Table 2.2; the corresponding standard deviations are listed in column 4 of Table 2.2. The values are resultant of two horizontal directions, as discussed in Chap. 1. The predicted values are highly uncertain because the standard deviations are nearly as high as the mean values. Unfortunately, the uncertainty in GMPMs has increased over the years [7]. The latest models are less certain and more complex than the older models. It is not reasonable for models to become less certain and more complex at the same time [7].

Table 2.1 Geological information about seismic sources (faults) in the region

Seismic sources	Magnitude	Distance	Occurrence rate	MRI
Fault 1	**M** 6	15 km	0.004/year	250 years
Fault 2	**M** 7.5	60 km	0.002/year	500 years

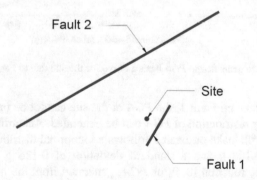

Fig. 2.1 Seismic sources (faults) that can produce earthquakes in the region surrounding the site

Table 2.2 Seismological information about ground motions produced by earthquakes

Earthquakes	Ground motion parameter	Mean	Standard deviation
M 6 at 15 km (Fault 1)	*PGA* (g)	0.168	0.126
	PGV (cm/s)	10.5	7.31
	PGD (cm)	1.65	1.64
M 7.5 at 60 km (Fault 2)	*PGA* (g)	0.103	0.0685
	PGV (cm/s)	10.3	6.91
	PGD (cm)	4.58	4.55

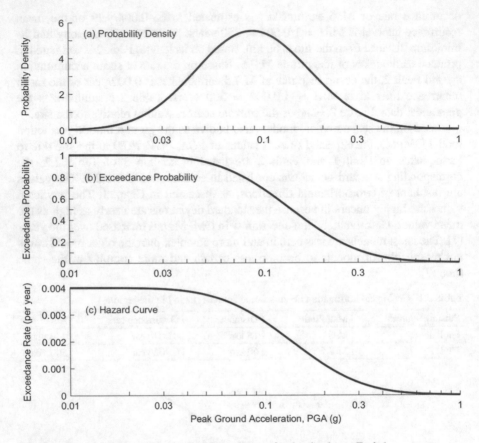

Fig. 2.2 Steps used to generate the *PGA* hazard curve for the site due to Fault 1

For an earthquake on Fault 1, the *PGA* at the site cannot be predicted precisely. Only a probability distribution of *PGA* can be generated. According to GMPM, the *PGA* due to an earthquake on Fault 1 follows a lognormal distribution [1, 2] with a mean value of 0.168 g and a standard deviation of 0.126 g (Table 2.2). The probability density function [8, 9] of *PGA*, generated from the mean and standard deviation of a lognormal distribution [8, 9], is shown in Fig. 2.2a. The exceedance probability function is one minus the cumulative distribution function [8, 9]; it is shown in Fig. 2.2b. According to GMPM, any value of *PGA* is possible if an earthquake occurs on Fault 1. Higher values of *PGA* have lower probabilities of exceedance and lower values of *PGA* have higher probabilities of exceedance. Naturally, the probability of *PGA* ≥ 0 is 1 (100%).

Since earthquakes on Fault 1 occur at a rate of 0.004/year, the exceedance probabilities in Fig. 2.2b are multiplied by 0.004 to obtain the exceedance rates for different values of *PGA*. These are shown in Fig. 2.2c. The plot in Figure 2.2c is known as the *PGA* hazard curve for the site due to Fault 1. The *PGA* hazard curve for the site due to Fault 2 is similarly generated, as shown in Fig. 2.3. Next, it is assumed

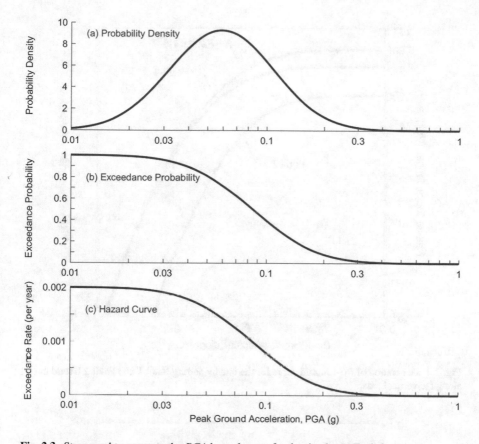

Fig. 2.3 Steps used to generate the *PGA* hazard curve for the site due to Fault 2

that: (1) earthquakes on Fault 1 and Fault 2 occur independently of each other and
(2) occurrence of an earthquake does not affect the chances of future earthquakes.
Therefore, the hazard curves due to Fault 1 and Fault 2 (Figs. 2.2c and 2.3c) can be
added along the vertical axis to obtain the overall *PGA* hazard curve for the site; it is
shown in Fig. 2.4. Note that the vertical scale in Fig. 2.4 is logarithmic while it was
linear in Figs. 2.2c and 2.3c.

The hazard curves for *PGV* and *PGD* are similarly generated; they are shown in
Figs. 2.5 and 2.6, respectively. For a time-independent process, the mean return
period (MRP) of exceedance is reciprocal of the rate of exceedance λ [8, 9]:

$$MRP = 1/\lambda \tag{2.1}$$

In Figs. 2.4, 2.5, and 2.6, the MRP of exceedance is shown on the right side. Note
that there is no upper limit on the values of *PGA*, *PGV*, and *PGD*. They increase with
increase in the MRP. Note also that Fault 1 controls the *PGA* at the site (Fig. 2.4) and
Fault 2 controls the *PGD* at the site (Fig. 2.6). Recall from Chap. 1 that *PGA* is a

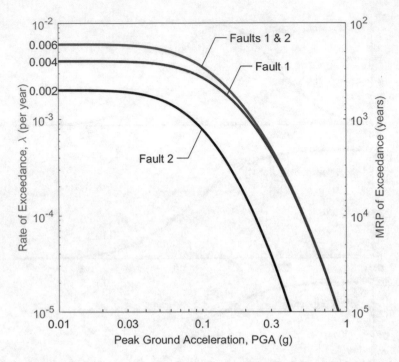

Fig. 2.4 Generation of *PGA* hazard curve for the site by adding Fault 1 and Fault 2 hazard curves along the vertical axis

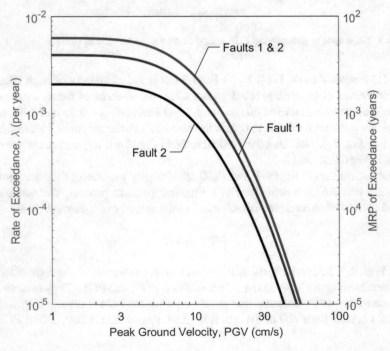

Fig. 2.5 Generation of *PGV* hazard curve for the site by adding Fault 1 and Fault 2 hazard curves along the vertical axis

measure of high frequencies in the ground motion and *PGD* is a measure of low frequencies in the ground motion. High frequencies cannot travel very far because of their short wavelengths, but low frequencies can travel far because of their long wavelengths. In general, closer seismic sources control *PGA* at the site even when they produce small earthquakes, and bigger seismic sources control *PGD* at the site even when they are far from the site. The PSHA is complete with the generation of *PGA*, *PGV*, and *PGD* hazard curves for the site.

2.2.2 Probability of Exceedance

For a time-independent process, the probability of exceedance within a given span τ is given by the following expression [8, 9]:

$$P = 1 - \exp\left(-\lambda\tau\right) \tag{2.2a}$$

or

$$P = 1 - \exp\left(-\frac{\tau}{MRP}\right) \tag{2.2b}$$

For 10% chance in 50 years, $P = 0.1$ (10%), and $\tau = 50$ years. Substituting these into Eq. 2.2b, the mean return period can be calculated as $MRP = 475$ years. Therefore, 10% chance of exceedance in 50 years is same as 475 years *MRP*. Table 2.3 presents MRP for some commonly used probabilities of exceedance.

2.2.3 Example: PSHA of a Real Site

For real sites, hundreds of seismic sources are used to perform PSHA. Some of these seismic sources are "faults" where earthquakes are known to have occurred in the past and can occur again in the future. Most of the seismic sources are polygons (areas) on the surface of the earth within which earthquakes may or may not have occurred in the known past. Within an *area source*, the chance of occurrence of an earthquake is considered uniform throughout. The site itself belongs to one of the area sources. The occurrence rate of an earthquake at a specific location within an area source is very small, but collectively area sources can have a significant influence on the seismic hazard at a site.

A site is selected in the San Francisco Bay Area (SFBA). Hundreds of hazard curves (corresponding to hundreds of seismic sources) are added together to obtain the overall hazard curves for the site. The *PGA*, *PGV*, and *PGD* values for various MRP are read from the hazard curves. They are listed in Table 2.4. Each set of *PGA*, *PGV*, and *PGD* values is used to generate a smooth response spectrum as described

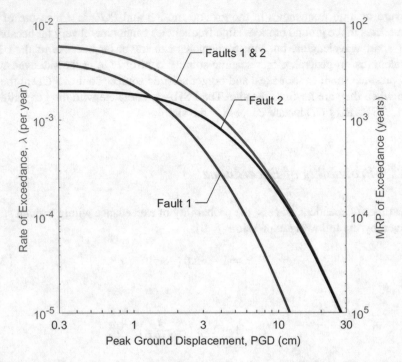

Fig. 2.6 Generation of *PGD* hazard curve for the site by adding Fault 1 and Fault 2 hazard curves along the vertical axis

Table 2.3 *MRP* for different probabilities of exceedance

50-year probability (%)	MRP (years)
50	72
20	224
10	475
5	975
2	2475
1	4975
0.5	9975

in Sects. 1.11 and 1.12. Figure 2.7 shows the 5% damping response spectra for the site for various MRPs.

2.3 Response Spectra of Vertical Ground Motion

Although vertical ground motion is not as important as the horizontal ground motion, it is still used in the analysis of some structures. The amplitudes of vertical ground motion are usually determined from those of the horizontal ground motion.

Table 2.4 *PGA*, *PGV*, and *PGD* for various MRP for the site in the SFBA

MRP (years)	*PGA* (g)	*PGV* (cm/s)	*PGD* (cm)
250	0.605	84.5	53.9
500	0.772	115	76.6
1000	0.915	139	103
2500	1.2	195	143
10,000	1.62	282	219

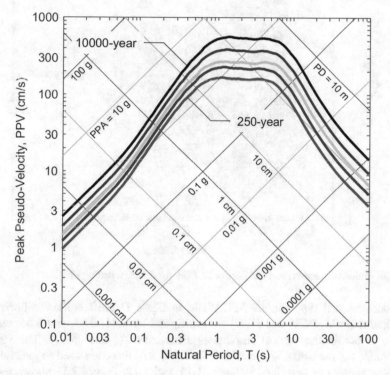

Fig. 2.7 Five-percent damping response spectra of horizontal ground motion for MRP of 250, 500, 1000, 2500 and 10,000 years for the SFBA site

The vertical to horizontal ratio V/H is most affected by the local soil conditions, because the soils are much more flexible in the horizontal direction than in the vertical direction. Figures 2.8, 2.9, and 2.10 show the ratios between the vertical and horizontal amplitudes for thousands of ground motions on different soil types. The median values are indicated by the red lines. The vertical to horizontal V/H ratios tend to be smaller for softer soils, because softer soils amplify horizontal motion more than vertical motion. The V/H ratio is smallest for *PGV* because horizontal *PGV* is most amplified by soils. Table 2.5 lists the median ratios between vertical and horizontal amplitudes for different soil types. These ratios are used to calculate the amplitudes of vertical ground motion from those of horizontal ground motion.

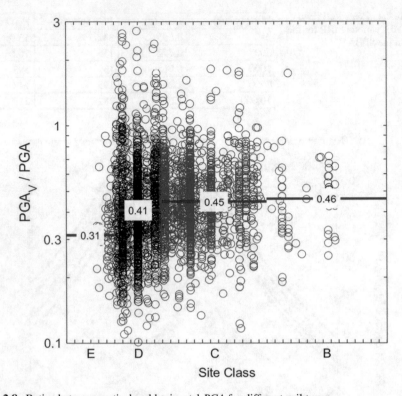

Fig. 2.8 Ratios between vertical and horizontal *PGA* for different soil types

Since the soil type for the SFBA site is Class D (stiff soil), the horizontal amplitudes (*PGA*, *PGV*, and *PGD*) are multiplied by 0.41, 0.35, and 0.35, respectively, to obtain the vertical amplitudes PGA_V, PGV_V, and PGD_V. The vertical amplitudes for the SFBA site are listed in Table 2.6, these are used to generate the response spectra as described in Sects. 1.11 and 1.12. Figure 2.11 shows the 5% damping response spectra of vertical ground motion for various MRP.

2.4   Building Code Response Spectra and Expected Performance

The design response spectra in building codes [10] are not purely probabilistic; they are adjusted by various factors (such as the "risk coefficients", "deterministic limit" and "2/3" factor in ASCE 7) to achieve consensus among various stakeholders in code committees. As a result of these artificial adjustments, the risk to code-designed structures is not the same everywhere [11, 12]. Building code response spectra are generated from three parameters S_s, S_1, and T_L [10], which are related to *PGA*, *PGV*, and *PGD*. It is not enough to select a response spectrum for design, it is equally

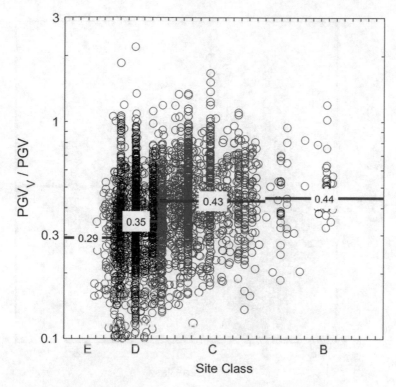

Fig. 2.9 Ratios between vertical and horizontal *PGV* for different soil types

important to select the desired performance objective. For example, a structure may be designed to remain operational (undamaged) during the 500-year MRP ground motion and to remain standing (without collapse) during the 2500-year MRP ground motion.

Structures cannot be made earthquake-proof because geoscientists are not able to provide an upper limit on the amplitude of ground motion that is possible at a site during the life of a structure. Even if such a limit existed, it will be prohibitively expensive to build earthquake-proof structures. In other words, the seismic risk to structures cannot be eliminated; it can only be reduced to an acceptable level. The MRP of design ground motion depends on the consequence of failure of the structure. For major dams and nuclear power plants, the consequence of failure is high. Therefore, the MRP of design ground motions is long. The consequence of collapse is much higher than the consequence of business interruption. Therefore, the MRP of collapse is much longer than the MRP of business interruption. Table 2.7 lists the approximate MRP of design ground motions for certain types of structures in the United States. Codes and regulations do not clearly define the MRP of operational performance and collapse prevention. They tend to be vague about the tolerable level of risk [11]. This is a shortcoming of codes and standards because the acceptable level of risk should be clearly communicated.

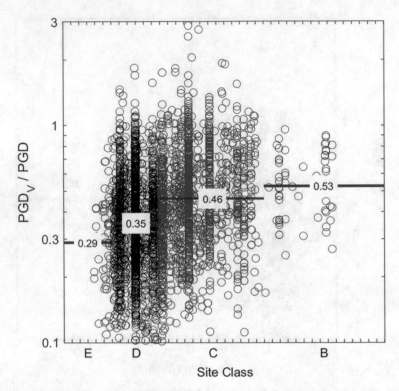

Fig. 2.10 Ratios between vertical and horizontal *PGD* for different soil types

Table 2.5 Median ratios between vertical and horizontal amplitudes for different soil types

Site class	PGA_V/PGA	PGV_V/PGV	PGD_V/PGD
A and B	0.46	0.44	0.53
C	0.45	0.43	0.46
D	0.41	0.35	0.35
E	0.31	0.29	0.29

Table 2.6 PGA_V, PGV_V, and PGD_V for various MRP for the site in the SFBA

MRP (years)	PGA_V (g)	PGV_V (cm/s)	PGD_V (cm)
250	0.248	29.6	18.9
500	0.316	40.2	26.8
1000	0.375	48.7	35.9
2500	0.491	68.1	50.2
10,000	0.662	98.8	76.6

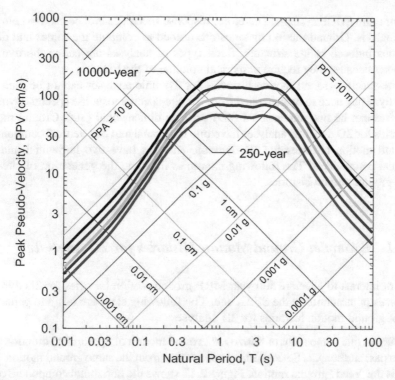

Fig. 2.11 Five-percent damping response spectra of vertical ground motion for MRP of 250, 500, 1000, 2500, 5000 and 10,000 years for the SFBA site

Table 2.7 Approximate MRP of design ground motions for different structures

MRP (years)	Immediate operational	Collapse/breach prevention
100	Nuclear power plants	
500	LNG facilities, existing hospitals, and schools	
1000	Major bridges, dams, hospitals, and schools	Existing hospitals, and schools
2500		Ordinary structures, LNG facilities
5000		Bridges, new hospitals, and schools
10,000		Major dams, nuclear power plants

2.5   Ground Motion Histories for Dynamic Analyses

It is usually possible to model a structure by one or more SDOF systems, whose responses can be read directly from the response spectra of ground motions. Such types of analyses are called static analyses because they do not require a numerical solution of the EOM. Sometimes, it is not possible to reliably model a structure by one or more SDOF systems. Therefore, response spectra cannot be used to complete the analysis. Ground motion histories are needed to compute the forces and deformations induced in the structure. Such types of analyses are called the dynamic analyses because they require numerical solution of the EOM.

Site-specific ground motion histories for dynamic analyses cannot be predicted directly. They need to be generated from the site-specific response spectra. Dynamic analyses can be two dimensional (2D) or three dimensional (3D). Ground motion histories for 2D dynamic analyses have one horizontal and one vertical components. Ground motion histories for 3D dynamic analyses have two horizontal and one vertical components. The following examples illustrate the generation of site-specific ground motion histories.

2.5.1 Example: Ground Motion Histories for 2D Analyses

It is of interest to generate 500-year MRP ground motion histories for 2D dynamic analyses of structures at the SFBA site. The following steps are taken to generate a set of ground motion histories for 2D analyses:

1. Select one component of horizontal ground motion of appropriate duration from a past earthquake. Select vertical component from the same ground motion. This is the "seed" ground motion. Figure 2.12 shows the horizontal component of the seed ground motion.
2. Generate the 5% damping response spectrum of the seed ground motion (Fig. 2.12) and compare it with the target (500-year MRP) response spectrum. The comparison is shown in Fig. 2.13.
3. Calculate ratios between the target response spectrum and the response spectrum of the seed ground motion for various periods. The ratios are shown in Fig. 2.14.
4. Generate Fourier amplitude and phase spectra of the seed ground motion using the *fft* routine in MATLAB [13]. The Fourier amplitude spectrum is shown by the blue curve in Fig. 2.15.
5. Multiply the Fourier amplitudes by the spectral ratios computed in Step 3. The modified Fourier amplitude spectrum of the seed ground motion is shown by the red curve in Fig. 2.15. The phase spectrum is not modified.

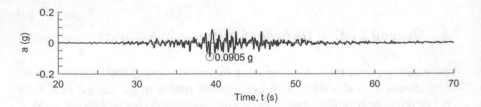

Fig. 2.12 Seed ground motion

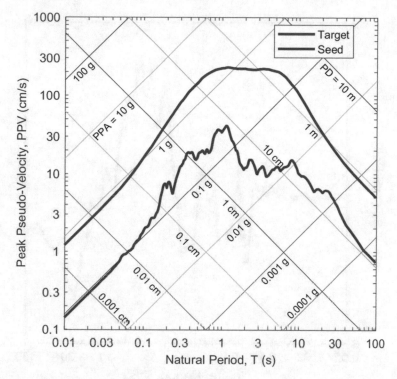

Fig. 2.13 Five-percent damping response spectrum of "seed" horizontal ground motion compared with the target (500-year MRP) response spectrum of horizontal motion

6. Combine the modified Fourier amplitude spectrum and the original phase spectrum to generate the modified seed ground motion using the inverse-Fourier *ifft* routine in MATLAB [13]. It is shown in Fig. 2.16.
7. Compute the response spectrum of the modified seed ground motion and compare it with the target response spectrum. The comparison is shown in Fig. 2.17.
8. Go to Step 9 if the spectrum of the modified seed ground motion satisfactorily matches the target response spectrum. Otherwise, replace the seed ground motion with the modified seed ground motion and repeat steps 2–7.
9. Repeat steps 2–8 to generate a site-specific vertical ground motion. Figure 2.18 shows a set of site-specific horizontal and vertical ground motion histories for 2D analyses.
10. Repeat steps 1–9 to generate seven sets of site-specific ground motion histories. Figure 2.19 compares the response spectra of seven site-specific horizontal ground motion histories with the target response spectrum of horizontal motion. The median of these seven spectra matches the target spectrum almost perfectly. Figure 2.20 compares the response spectra of seven site-specific vertical ground motion histories with the target response spectrum of vertical motion. Again, the median of these seven spectra matches the target spectrum almost perfectly.

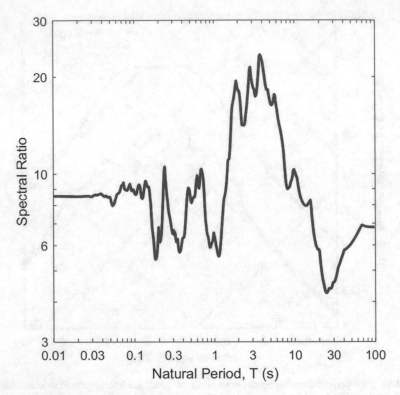

Fig. 2.14 Ratio between the target response spectrum and the spectrum of seed horizontal ground motion shown in Fig. 2.13

2.5.2 Example: Ground Motion Histories for 3D Analyses

3D analyses are carried out with two horizontal and one vertical ground motion histories. The following steps are taken to generate a set of ground motion histories for 3D analyses:

1. Select all three components of a "seed" ground motion of appropriate duration from a past earthquake.
2. Compute the resultant response spectrum of two horizontal components and compare it with target response spectrum of horizontal ground motion.
3. Compute the ratios between the target response spectrum and the resultant horizontal response spectrum of seed ground motion.
4. Generate Fourier amplitude and phase spectra of the seed ground motion components.
5. Multiply the Fourier amplitude spectra of both horizontal components with the spectral ratios computed in Step 3 to compute the modified Fourier amplitude spectra. The phase spectra of both horizontal components are not modified.

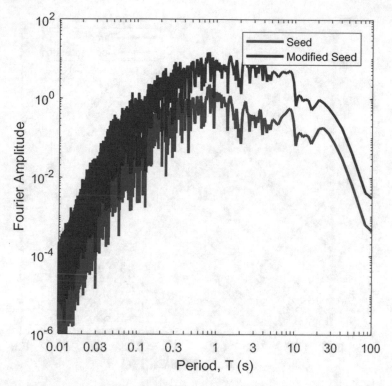

Fig. 2.15 Fourier amplitude spectra of seed horizontal ground motion and modified seed horizontal ground motion

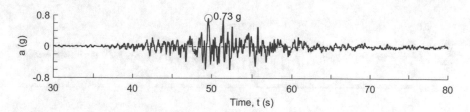

Fig. 2.16 Modified seed ground motion

6. Combine the modified Fourier amplitude spectra with the original phase spectra of seed ground motion components to compute the modified ground motion components.
7. Compute the resultant response spectrum of modified ground motion components and compare it with the target response spectrum.
8. Go to Step 9 if the spectrum of modified ground motion components satisfactorily matches the target response spectrum. Otherwise, replace the seed ground motion components with the modified seed ground motion components and repeat steps 2–7.

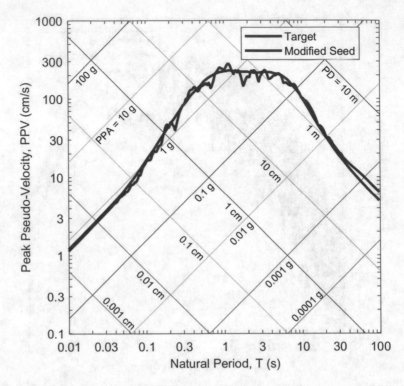

Fig. 2.17 Response spectrum of modified seed horizontal ground motion compared with the target response spectrum of horizontal motion

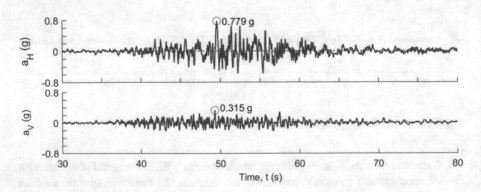

Fig. 2.18 One set of horizontal and vertical ground motions for 2D analyses

9. Generate vertical ground motion history similar to that for the 2D analyses.
10. Repeat steps 1–9 to generate seven sets of site-specific ground motions for 3D analyses. Figure 2.21 shows one set of ground motions for 3D analyses.

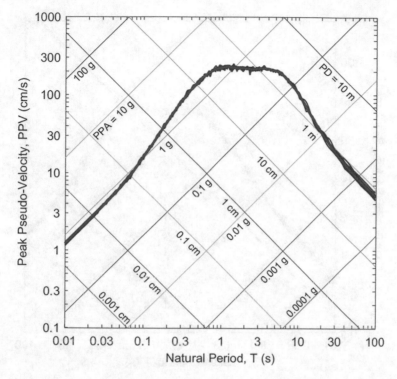

Fig. 2.19 Response spectra of seven site-specific horizontal ground motions compared with the target spectrum of horizontal motion

2.5.3   Unequal Ground Motions in Two Horizontal Directions

Seismic hazard at a site is usually determined by numerous seismic sources which are at various distances from the site and various orientations with respect to the site. Therefore, it is not possible to say in which horizontal direction the shaking will be stronger. However, in rare cases, the hazard may be controlled by a single seismic source close to the site and it may be possible to predict the horizontal direction of stronger shaking. If that is the case, the seed ground motions with significantly different horizontal components can be selected, but the resultant of those two components should still be matched with the site-specific response spectrum.

2.6   Response Spectra for Various Values of Damping

For many applications, response spectra are needed for damping other than 5% of critical. Once spectrum-compatible ground motion histories have been generated, response spectra for many different values of damping can be generated. Figure 2.22

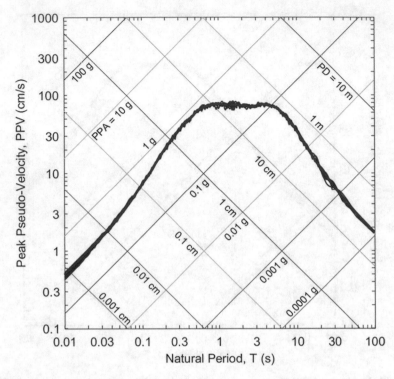

Fig. 2.20 Response spectra of seven site-specific vertical ground motions compared with the target spectrum of vertical motion

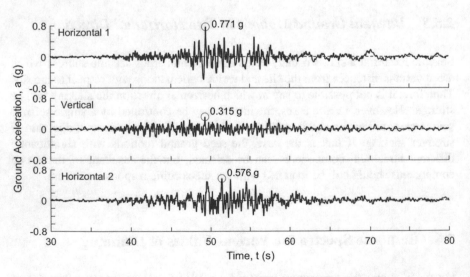

Fig. 2.21 One set of horizontal and vertical ground motion histories for 3D analyses

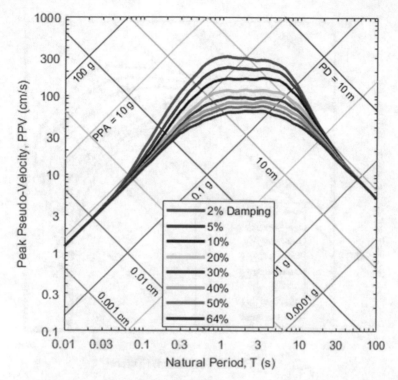

Fig. 2.22 500-year MRP response spectra for the SFBA site for different values of damping

shows a set of 500-year MRP response spectra for different values of damping. These are calculated directly from the seven sets of site-specific ground motion histories. For each damping, seven different response spectra were obtained. These were averaged and smoothed to generate the plots in Fig. 2.22. Figure 2.23 shows the 500-year MRP ADRS for the SFBA site for various values of damping. Figure 2.24 shows linear plots of the 500-year MRP ADRS for various values of damping.

The ADRS plots show the strength- and deformation demands imposed by the 500-year MRP ground motion on structures of various values of damping. There are different ways to interpret the ADRS; one useful way is to read the deformation demands for various values of strength and damping. Table 2.8 lists the deformation demands for selected values of strength and damping. For the same damping, the deformation demand reduces with increase in strength. For the same strength, the deformation demand reduces with increase in damping. A structure with high deformability and/or high damping need not be very strong to resist seismic ground shaking. The deformability and damping of a structure can be increased by thoughtful detailing.

Important sources of deformability are:

1. Elastic deformation
2. Plastic deformation (yielding)

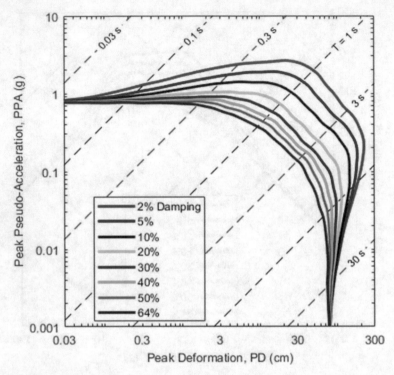

Fig. 2.23 ADRS plots of 500-year MRP response spectra for the SFBA site for different values of damping

3. Sliding at the base
4. Rocking (uplifting) at the base
5. Deformation of soil below the foundation

Important sources of damping are:

1. Material damping during elastic deformation
2. Hysteretic damping due to plastic yielding
3. Sliding at the base
4. Base impacts during rocking
5. Soil damping
6. Radiation of waves away from the foundation

In a good seismic analysis, all sources of deformability and damping are explicitly considered.

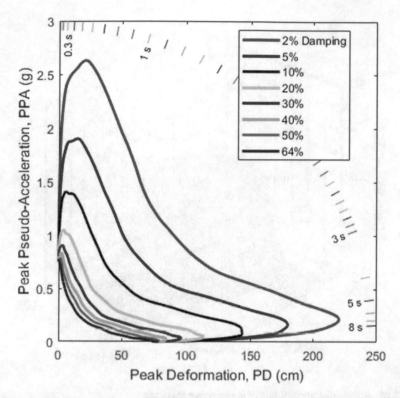

Fig. 2.24 Linear plots of 500-year MRP ADRS for the SFBA site for different values of damping

Table 2.8 Deformation demands on structures of various strength and damping

Lateral strength (% weight)	Deformation demand (cm)			
	$\zeta = 5\%$	$\zeta = 10\%$	$\zeta = 20\%$	$\zeta = 30\%$
20	178	127	68	45
30	150	95	45	30
40	122	67	35	23
50	94	55	27	18
60	79	46	23	13

2.7 Demand Surface

ADRS for various values of damping (Fig. 2.24) can be stacked on top of each other to generate a three-dimensional surface, as shown in Fig. 2.25. This is a complete description of strength and deformation demands on structures of various damping. This plot will be helpful in later chapters to visualize the seismic response of structures whose strength and damping depend on deformation.

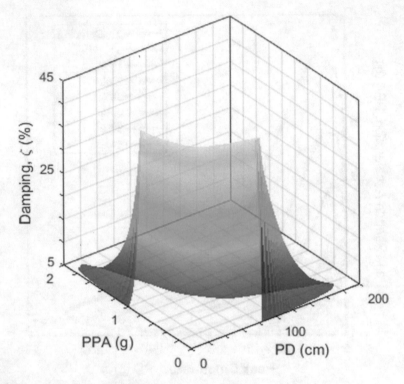

Fig. 2.25 A three-dimensional plot of the response spectrum

2.8 Energy Demand for SFBA Site

As discussed in the previous chapter, there is no single parameter which completely defines the seismic demand imposed by the ground motion, but the area enclosed within the 5% damping ADRS comes close to defining the seismic demand for a broad range of structures. As shown in Fig. 2.26, $ED = 13.4$ (m/s)^2 for the 500-year MRP ground motion at the SFBA site. ED has the units of energy/mass. ED can also be computed for different MRP and different damping. Table 2.9 lists ED values for different damping and different MRP. With tenfold increase in damping from 5 to 50%, ED reduces to one-eighths. With tenfold increase in MRP from 500- to 5000-year, ED increases to four times.

2.9 Structure Types for Different Sites

Certain types of structures are better suited for a specific site. On a rock site, high-frequency ground motions are expected during closer–small earthquakes. Figure 2.27 shows ADRS of a high-frequency ground motion. Notice that the strength demand is

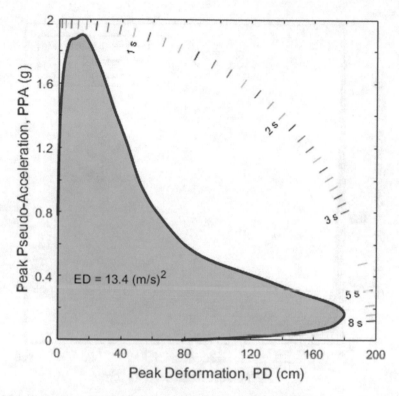

Fig. 2.26 Five-percent damping 500-year MRP ADRS for the SFBA site

Table 2.9 *ED* for SFBA site for different damping and different MRP

MRP	Energy demand *ED* (m/s)^2					
	$\zeta = 5\%$	$\zeta = 10\%$	$\zeta = 20\%$	$\zeta = 30\%$	$\zeta = 50\%$	$\zeta = 64\%$
250	7.3	4.2	2.2	1.5	0.9	0.7
500	13.4	7.6	4	2.7	1.7	1.2
1000	19.7	11.2	5.9	4	2.5	1.8
2500	38.4	21.9	11.5	7.7	4.9	3.4
5000	53.6	30.6	16	10.8	6.8	4.8
10,000	80.8	46.1	24	16.3	10.3	7.2

high, but the deformation demand is low. Therefore, deformable structures, such as mid- to high-rise buildings, are better suited for this site.

On soil sites, low-frequency ground motions are expected during distant–larger earthquakes. Figure 2.28 shows ADRS of a low-frequency ground motion. Notice that the deformation demand is high, but the strength demand is low. Therefore, stiff and strong structures, such as low-rise buildings, are better suited for this site.

On "soft-soil" sites with well-defined natural frequency, narrow-banded ground motions are expected. Figure 2.29 shows ADRS of a narrow-banded ground motion. Notice the effect of damping on strength and deformation demands. Heavily damped structures are most suited for this site.

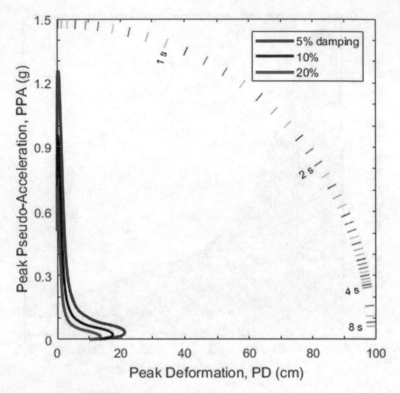

Fig. 2.27 ADRS of high-frequency ground motion expected on a rock site during closer–small earthquakes

2.10 Summary

1. Due to high uncertainty in geological and seismological models, probabilistic analyses are needed to predict future ground motions.
2. There is no upper limit on the intensity of ground shaking that is possible at a site. Therefore, the design ground motions have a certain chance of being exceeded during the life of a structure.
3. Closer seismic sources control *PGA* at the site even when they produce small earthquakes. Bigger seismic sources control *PGD* at the site even when they are far from the site.
4. Site-specific design ground motions are expressed in two different ways: (1) a set of response spectra for various values of damping and (2) a set of spectrum-compatible ground motion histories.
5. Response spectra are used in static analyses. Ground motion histories are used in dynamic analyses.
6. Structures with high consequence of failure are designed to less likely (or longer MRP) ground motions. Structures with low consequence of failure are designed to more likely (or shorter MRP) ground motions.

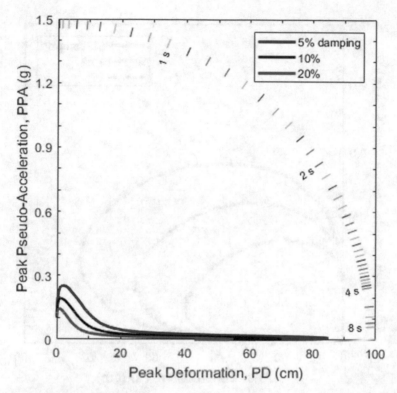

Fig. 2.28 ADRS of low-frequency ground motion expected on a soil site during distant–large earthquakes

7. For the same structure, the consequence of collapse is much higher than the consequence of business interruption. Therefore, the design ground motions for collapse prevention have longer MRP than design ground motions for business continuity.
8. Cities, states, and federal governments are more focused on collapse prevention. Owners, operators, and insurance companies are also interested in business continuity.
9. For the same damping, increase in deformability reduces the strength demand on structures.
10. For the same deformability, increase in damping reduces the strength demand on structures.
11. Deformability and damping can be increased by thoughtful detailing of the structure.
12. Deformable structures are most effective against high-frequency ground motions; strong structures are most effective against low-frequency ground motions; and damped structures are most effective against narrow-banded ground motions.

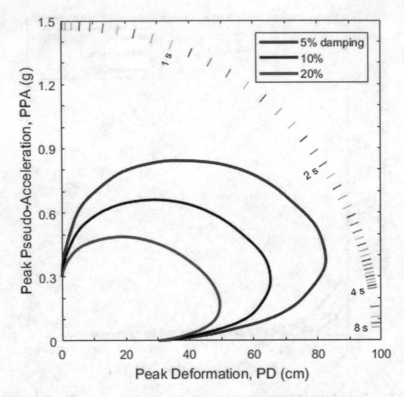

Fig. 2.29   ADRS of narrow-banded ground motion expected on a "soft-soil" site

References

1. Esteva, L. (1970). Seismic risk and seismic design decisions. In R. J. Hansen (Ed.), *Seismic design of nuclear power plants*. Cambridge, MA: MIT Press.
2. Cornell, C. A. (1971). Probabilistic analysis of damage to structures under seismic loads. In D. A. Howells, P. Haigh, & C. Taylor (Eds.), *Dynamic waves in civil engineering* (pp. 473–488). New York: Wiley.
3. Reiter, L. (1991). *Earthquake hazard analysis—Issues and insights* (p. 254). New York: Columbia University Press.
4. McGuire, R. K. (2004). *Seismic hazard and risk analysis* (p. 221). Oakland, CA: Earthquake Engineering Research Institute.
5. Petersen, M. D., Moschetti, M. P., Powers, P. M., Mueller, C. S., Haller, K. M., Frankel, A. D., et al. (2014). *Documentation for the 2014 Update of the United States National Seismic Hazard Maps*. U.S. Geological Survey. Open-File Report 2014–1091, 243pp.
6. FEMA (2003). *NEHRP recommended provisions and commentary for seismic regulations for new buildings and other structures* (2003 ed.). Federal Emergency Management Administration (FEMA).
7. Malhotra, P. K. (2015, July). Myth of probabilistic seismic hazard analysis. *Structure Magazine*, pp. 58–59.
8. Ang, A. H.-S., & Tang, W. H. (1975). *Probability concepts in engineering planning and design* (p. 1). New York: Basic Principles, John Wiley.

9. Nowak, A. S., & Collins, K. R. (2000). *Reliability of structures* (p. 2000). New York, NY: McGraw-Hill.
10. ASCE. (2016). *Minimum design loads for building and other structures* (ASCE Standard No. ASCE 7–16). Reston, VA: American Society of Civil Engineers.
11. Malhotra, P. K. (2005). Return period of design ground motions. *Seismological Research Letters, 76*(6), 693–699.
12. Malhotra, P. K. (2006). Seismic risk and design loads. *Earthquake Spectra, 22*(1), 115–128.
13. MathWorks. (2020). *MATLAB Version 9.8.0.1417392 (R2020a)*. Natick, MA: MathWorks.

Chapter 3
Seismic Response of One-Story Buildings

Nomenclature

ζ	Viscous damping ratio
2D	Two dimensional
ADRS	Acceleration–deformation response spectrum
EBF	Eccentric braced frame
EOM	Equation(s) of motion
g	$9.81 \text{ m/s}^2 =$ acceleration due to gravity
MF	Moment frame(s)
MRP	Mean return period
PD	Peak deformation
PPA	Peak pseudo-acceleration
SDOF	Single degree of freedom system
SFBA	San Francisco Bay Area
ST	Seismic toughness (area below the capacity curve)
T	Natural period of structure

3.1 Introduction

In Chap. 2, ground motions for future earthquakes were generated for a site in the San Francisco Bay Area (SFBA). Ground motions were expressed in two different ways: (1) a set of response spectra for various values of damping and (2) a set of spectrum-compatible ground motion histories. Key characteristics of a structure are its mass (weight), lateral strength, deformability, and damping. Lateral strength is structure's ability to resist horizontal load. Deformability is its ability to deform laterally. Damping is structure's ability to quickly dissipate vibration energy. For a specific damping, a more deformable structure needs smaller strength to withstand

© Springer Nature Switzerland AG 2021
P. K. Malhotra, *Seismic Analysis of Structures and Equipment*,
https://doi.org/10.1007/978-3-030-57858-9_3

ground shaking. For a specific deformability, a more damped structure needs smaller strength to withstand ground shaking. The lateral strength of a structure depends on: (1) the size, configuration, and material of structural elements (beams, columns, braces, etc.), (2) the strength of connections, and (3) the strength of the foundation. The deformability and damping of a structure can be enhanced by proper detailing—by encouraging ductile modes of failure (such as flexural yielding in beams), and suppressing brittle modes of failure (such as shear failure in concrete beams and columns, buckling in steel columns and braces, and failure of connections).

There are different ways of analyzing the response of a structure to ground motions. The analyses can be static or dynamic. Static analyses do not require the solution of the equations of (EOM); they use the response spectra for various values of damping. Dynamic analyses require the solution of the EOM; they use ground motion histories. The analyses can be linear or nonlinear. Linear analyses assume that the force induced in a structure is linearly proportional to its deformation—doubling the deformation doubles the force. Nonlinear analyses assume a more realistic relationship between the force and deformation. There are four basic types of analyses: (1) linear-static, (2) linear-dynamic, (3) nonlinear-static, and (4) nonlinear-dynamic. In this chapter, pros and cons of different types of analyses are discussed for a one-story moment-frame (MF) building. The effect of braces is discussed near the end of the chapter.

3.2 One-Story Moment Frame

Figure 3.1 shows the sketch of a one-story steel MF fabricated from W14 × 109 columns and W14 × 99 beams. The steel type is ASTM A36 with an expected yield strength of 250 MPa. Column-foundation and column-beam connections are rigid (fully restrained). During horizontal shaking of the ground, columns and beam deform due to bending, as shown in Fig. 3.1. Rigid connections ensure that the angles between the columns and foundations and between columns and beam remain 90° even after the MF has deformed. If the connections are stronger than the connected members, plastic yielding will occur in the members before brittle failure of the connections. This is a highly desirable characteristic of a moment frame because it results in a ductile system which can deform and dissipate energy. Specific

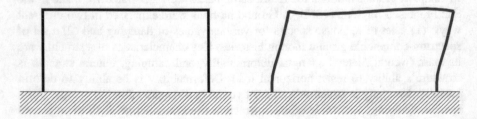

Fig. 3.1 Sketch of a one-story moment frame (MF)

Table 3.1 Important structural properties of the MF shown in Fig. 3.1

Length of beam (distance between column centerlines)	7.32 m
Height of column	3.66 m
Lumped mass	78.4×10^3 kg
Moment of inertia of column cross section	51,694 cm^4
Moment of inertia of beam cross section	46,274 cm^4
Plastic moment capacity of columns	781 kNm
Plastic moment capacity of beam	704 kNm
Maximum allowable plastic hinge rotation in columns	0.0568 radian [6]
Maximum allowable plastic hinge rotation in beam	0.0783 radian [6]

guidelines need to be followed to ensure ductile behavior of steel MF [1–3]. Sometimes, members are intentionally weakened to ensure that plastic hinges will form in the members before failure of the connections [4]. After yielding, plastic hinges appear at the base of columns and at the ends of beam where the moments are maximum. Due to plastic hinge rotations, the angles between the columns and foundations and between beam and columns are no longer 90°. They are greater than or smaller than 90° depending on the direction of plastic rotation. The maximum allowable plastic hinge rotation in column depends on the "compactness" of the cross section, slenderness ratio of the column in the weak direction, and the axial load [5, 6]. The maximum allowable plastic hinge rotation in beam depends on the "compactness" of the cross section and the length of beam [6].

Table 3.1 lists the important structural properties of the MF shown in Fig. 3.1. The mass of the structure lumped with the MF is $m = 78.4 \times 10^3$ kg. The mass is lumped at the beam level, and the beam is assumed axially rigid. Therefore, the MF can be treated as a single-degree-of-freedom (SDOF) system.

3.3 Ground Motion

In this chapter, the response of the MF is computed during the 500-year MRP ground motion at a site in the San Francisco Bay Area (SFBA). The 500-year MRP ground motion for the SFBA site was generated in Chap. 2. Figure 3.2 shows the 500-year MRP response spectra for various values of damping. The response spectra are shown in the acceleration–deformation format. These are known as the demand curves for the site. The peak pseudo-acceleration PPA is read along the vertical axis and the peak deformation PD is read along the horizontal axis. The natural period T is shown by the parallel diagonal lines. The PPA, PD, and T are related to each other by the expression: $PD = PPA \times (T/2\pi)^2$. Figure 3.3 shows one of the seven spectrum-compatible ground motion histories for 2D analyses; these were also generated in Chap. 2. Different types of analysis are discussed next.

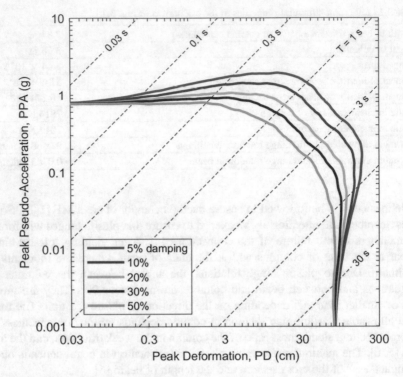

Fig. 3.2 500-year MRP ADRS for various values of damping for the SFBA site

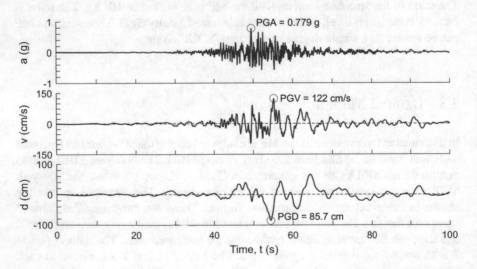

Fig. 3.3 One of seven 500-year MRP spectrum-compatible ground motion history

3.4 Linear-Static Analysis

The linear analysis is applicable as long as the MF does not yield or break anywhere. The lateral stiffness of the MF can be determined with the help of a structural analysis program such as SAP 2000 [7]; it is $k= 24.1 \times 10^6$ N/m. Therefore, the natural period of the system is $T = 2\pi\sqrt{m/k} = 0.358$ s. In a linear system, some sources of damping are air resistance, internal friction, damage to nonstructural systems such as walls and ceilings, and radiation of energy through the foundation. The first two sources of damping are usually small. For the MF, the radiation damping is also expected to be small because less deformation occurs in the foundation relative to the structure. Therefore, a nominal damping of 0.05 (5% of critical) is assumed for the linear analysis.

Figure 3.4 shows the pseudo-acceleration plot of the 5% damping 500-year MRP response spectrum for the SFBA site. The peak pseudo-acceleration for $T = 0.358$ s is $PPA = 1.83$ g. Thus, the peak deformation is $PD = PPA \times (T/2\pi)^2 = 5.85$ cm. Alternatively, both PPA and PD can be simultaneously read from the 500-year MRP ADRS shown in Fig. 3.5. The induced base shear (horizontal force) is $Q = m \cdot PPA = 78.4 \times 10^3 \times 1.83 \times 9.81 = 1.41$ MN. The same value of the base shear can be

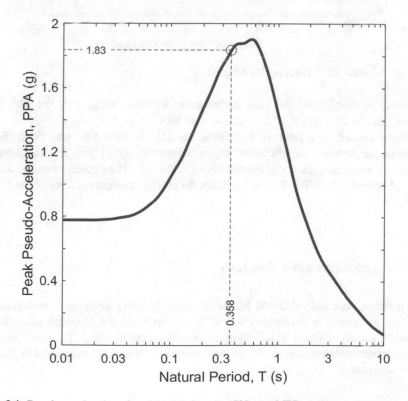

Fig. 3.4 Pseudo-acceleration plot of the 5% damping 500-year MRP response spectrum

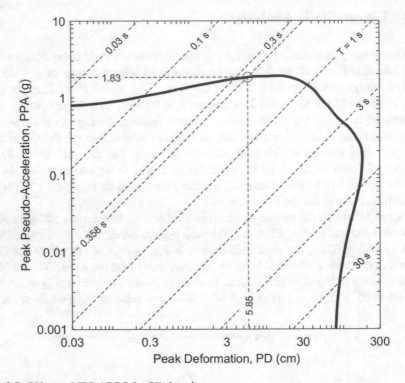

Fig. 3.5 500-year MRP ADRS for 5% damping

obtained by multiplying the peak deformation with the stiffness of the MF, i.e.,
$Q = k \cdot PD = 24.1 \times 10^6 \times 5.85/100 = 1.41$ MN.

When pushed by a force of 1.41 MN, the MF deforms 5.85 cm. From static analysis, the induced moment at the base of columns is 1653 kNm. This is 2.1 times the plastic moment capacity of the columns (Table 3.1). The induced moment at the ends of beam is 916 kNm. This is 1.3 times the plastic moment capacity of the beam (Table 3.1).

3.5   Linear-Dynamic Analysis

For a linear, viscously damped SDOF system, dynamic analysis is unnecessary because the response spectrum itself is generated from the dynamic analysis of linear, viscously damped SDOF systems, as discussed in Chap. 1. Therefore, the results of linear-dynamic analysis can be assumed to be same as the results of linear-static analysis.

3.6 Need for Nonlinear Analysis

According to the linear analysis, the moments in beam and columns exceed their plastic moment capacities. In other words, the MF lacks the strength to remain fully elastic during the 500-year MRP ground motion. If the connections are weaker than the connected elements, the connections will break in a brittle fashion and the MF could collapse during the 500-year MRP ground motion. If, on the other hand, the connections are stronger than the connected elements, the beam and columns will yield during the 500-year MRP ground motion. The linear analysis cannot predict the extent of plastic yielding. The plastic yielding in the MF is measured in terms of the plastic-hinge rotation. Strictly speaking, any plastic rotation will cause some structural damage to the MF and reduce its resistance to subsequent earthquakes. However, small plastic rotations (<0.01 radian) do negligible damage [6], hence they can be tolerated. Plastic rotations between 0.02 and 0.04 radian do nontrivial damage, hence they may require some repairs after the earthquake. Plastic rotations, approaching the maximum allowable values (Table 3.1), can cause breaks at the hinges, resulting in partial collapse of the MF. A nonlinear analysis is needed to determine the condition of the MF following the 500-year MRP ground motion at the site. Two methods of nonlinear analysis are discussed next.

3.7 Nonlinear-Static Analysis

The basis of nonlinear-static analysis is to determine the strength, deformability, and damping of the MF and compute its response without performing a dynamic analysis. The lateral strength and deformability of a system are expressed by its pushover curve. The damping of a nonlinear system is not fixed; it depends on deformation. A plot of damping versus deformation is known as the damping curve. The pushover and the damping curves for the MF are generated next.

3.7.1 Pushover Curve

The pushover curve is a plot between the lateral force and the lateral deformation of the system. The lumped mass, at the beam level, is gradually pushed in the horizontal direction, causing the bending moments to rise in the beam and the columns. At some point, plastic hinges appear at the base of columns. At some later point, plastic hinges also appear at the ends of beam. Figure 3.6 shows the forces and moments acting on columns after all the plastic hinges have formed in the MF. Disregarding the effect of gravity, the lateral force after the formation of all plastic hinges is given by the following expression:

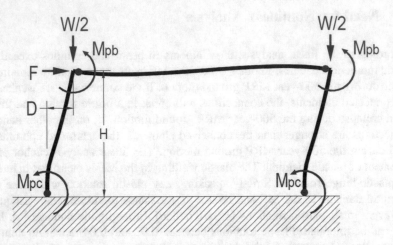

Fig. 3.6 Forces and moments on columns after the formation of all plastic hinges

$$F_p = \frac{2(M_{pb} + M_{pc})}{H} \tag{3.1}$$

in which M_{pb} = plastic moment capacity of the beam cross section; M_{pc} = plastic moment capacity of the column cross section; and H = height of the MF. Upon substituting M_{pb}, M_{pc}, and H from Table 3.1, F_p = 811 kN.

The elastic stiffness, before the formation of any plastic hinges, is given by the following expression:

$$k = m\left(\frac{2\pi}{T}\right)^2 \tag{3.2}$$

Substituting, $m = 78.4 \times 10^3$ kg and $T = 0.358$ s into Eq. 3.2 gives $k = 24.1$ MN/m. The maximum elastic deformation is:

$$D_e = \frac{F_p}{k} \tag{3.3}$$

Substituting $F_p = 811$ kN and $k = 24.1$ MN/m into Eq. 3.3 gives $D_e = 3.37$ cm. The elastic portion of the pushover curve ($D \leq D_e$) is generated by using the following relationship:

$$F = k \cdot D \tag{3.4}$$

For deformation $D \leq D_e$, the force increases linearly with D, but the plastic rotation θ remains zero. For $D > D_e$, the force remains fixed at $F = F_p$, but the plastic rotation θ increases with deformation. θ is the same for both beam and columns. The relationship between D and θ is as follows:

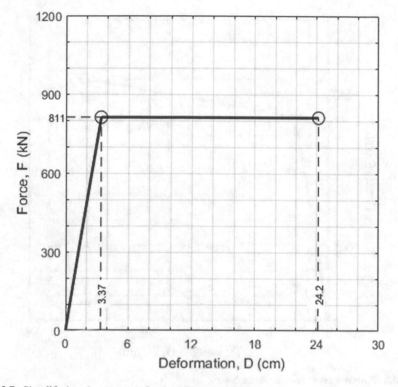

Fig. 3.7 Simplified pushover curve for the MF

$$D = D_e + \theta \cdot H \tag{3.5}$$

The maximum allowable plastic hinge rotation is 0.0568 radian for columns and 0.0783 radian for beam (Table 3.1). Since columns and beam experience the same plastic rotation, the allowable rotation in columns controls the deformation of the MF at collapse, i.e., $3.37 + 0.0568 \times H = 24.2$ cm. Figure 3.7 shows a simplified version of the pushover curve obtained from Eqs. 3.1–3.5. In a more accurate pushover curve, the transition from elastic to plastic is more gradual as plastic hinges appear one by one. The pushover curve of Fig. 3.7 ignores any increase in plastic moment with increase in rotation, due to strain hardening. Finally, the pushover curve ignores the effect of gravity, known as the P–Δ effect [8]. The pushover curve is refined as follows:

Strain hardening. ASCE 41 [6] recommends that the post-yield stiffness can be assumed equal to 3% of the elastic stiffness. This is somewhat arbitrary because strain hardening in hinges should not depend on the elastic deformation of the MF. But due to lack of better information, ASCE 41 [6] recommendation is followed. Figure 3.8 shows the pushover curve with strain hardening. The lateral strength of the MF increases from 811 to 960 kN. This is an 18% increase in lateral

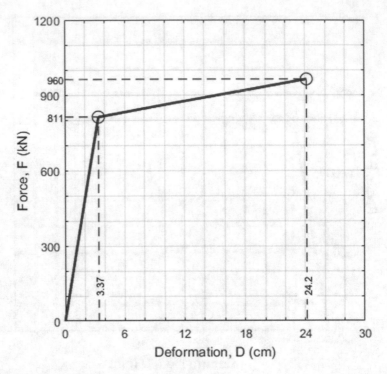

Fig. 3.8 Pushover curve with strain hardening

strength while the plastic rotation increases from 0 to 0.0568 radian. The ultimate moment capacity of columns is 18% higher than the value listed in Table 3.1. The ultimate moment capacity of beam is $18/0.0568 \times 0.0783 = 24.8\%$ higher than the value listed in Table 3.1. Therefore, the ultimate moment capacities of columns and beam are 922 and 879 kNm, respectively.

The ultimate strength of 910 kN is 13% higher than the yield strength of 804 kN. This increase occurs while the deformation D increases from 3.37 cm to 24.2 cm or the plastic rotation increases from 0 to 0.0568 radian. Therefore, the net increase in strength is $13/0.0568 = 230\%$ per radian.

P–Δ effect. Referring to Fig. 3.6, note that the weight of the structure applies an additional moment which is in the same direction as the moment applied by the lateral force F. Both these moments are resisted by the MF. The weight of the structure reduces the lateral force F for a given deflection D. P–Δ effect is considered as follows: (1) for a given value of D, F is read from Fig. 3.8 and (2) F is reduced by an amount DW/H, where $W =$ the weight of the structure. Figure 3.9 shows the refined pushover curve after considering both strain hardening and P–Δ effect. Due to P–Δ, the yield strength of the MF reduces from 811 kN to 804 kN, and the ultimate strength of the MF reduces from 960 kN to 910 kN.

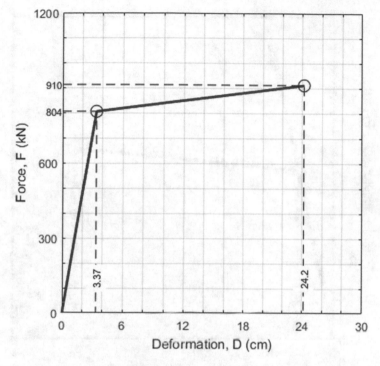

Fig. 3.9 Pushover curve with strain hardening and P–Δ effect

3.7.2 Capacity Curve

The pushover curve of Fig. 3.9 is converted into the capacity curve by dividing the force along the vertical axis of Fig. 3.9 by the mass $m = 78.4 \times 10^3$ kg. Figure 3.10 shows a plot of the capacity curve. The vertical axis in Fig. 3.10 is the pseudo-acceleration PA. Recall from Chap. 1 that the natural period depends on the ratio between the deformation and the pseudo-acceleration, i.e.,

$$T = 2\pi\sqrt{\frac{D}{PA}} \tag{3.6}$$

For a nonlinear system, it is more appropriate to refer to T as the "effective period."

With the help of Eq. 3.6, radial tick marks corresponding to various periods are drawn in Fig. 3.10. Note that the period of the MF remains fixed at 0.358 s up to the formation of plastic hinges at 3.37 cm deformation. After that, the period increases with increase in deformation. The capacity curve ends at 24.2 cm when the plastic rotation in columns reaches its maximum allowable value of 0.0568 radian. The area below the capacity curve (Fig. 3.11) is called the seismic toughness ST. In general,

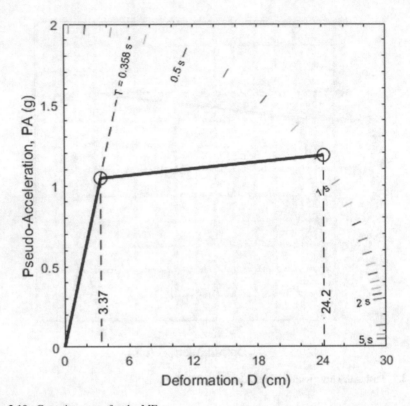

Fig. 3.10 Capacity curve for the MF

structures with greater toughness perform better during earthquakes. For the MF discussed here, $ST = 2.45$ (m/s)^2. ST has the units of energy/mass.

In Fig. 3.12, the capacity curve of Fig. 3.10 is redrawn on a logarithmic scale. The radial tick marks in Fig. 3.10 are replaced by parallel diagonal lines in Fig. 3.12. Both linear and logarithmic plots of the capacity curve (Figs. 3.10 and 3.12) will be used in the nonlinear-static analysis.

3.7.3 Damping Curve

During earthquake, the structure deforms back and forth in a cyclic manner. Figure 3.13 shows the cyclic force–deformation relationship for the MF subjected to deformation cycles of various amplitudes. It is assumed that there is no degradation in strength, stiffness, and damping under repeated cycles up to the limiting plastic rotations recommended in ASCE 41 [6]. The force–deformation relationship is hysteretic— meaning that some energy is lost during each cycle. The energy loss occurs due to plastic-hinge rotations and it equals the area enclosed within the force–deformation loop. This is the primary source of damping in a moment frame. The damping depends on deformation. The damping for various values of deformation is determined next.

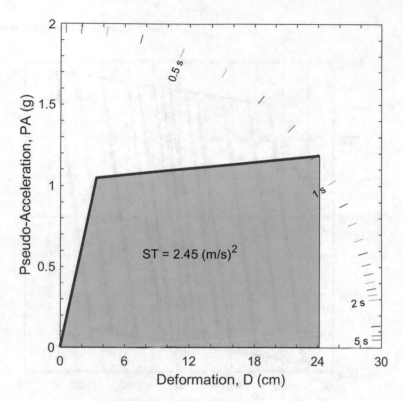

Fig. 3.11 Seismic toughness of the MF

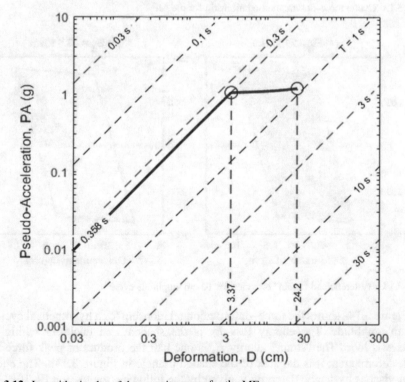

Fig. 3.12 Logarithmic plot of the capacity curve for the MF

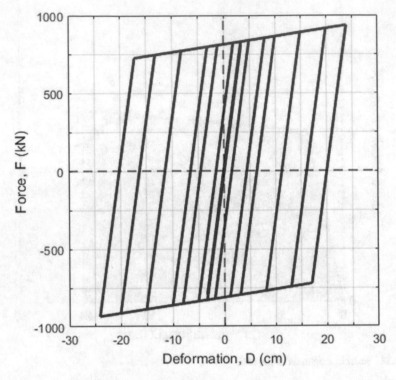

Fig. 3.13 Cyclic force–deformation relationship for the MF

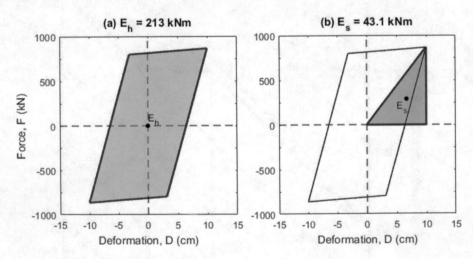

Fig. 3.14 Hysteretic and "strain" energies for 10-cm amplitude cycle

Figure 3.14a shows the force–deformation relationship for a hypothetical cycle of 10-cm amplitude. The energy loss E_h is the shaded area enclosed within the hysteresis loop. The "strain" energy E_s is one-half the product of peak force and peak deformation; it is the area of the shaded triangle in Figure 3.14b. The equivalent-viscous hysteretic damping is given by the following expression [9, 10]:

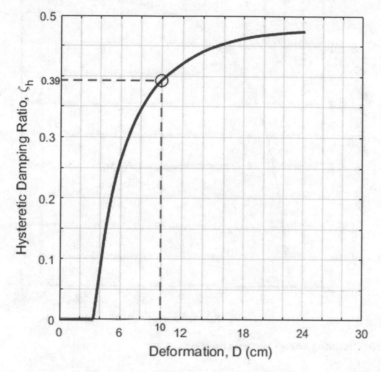

Fig. 3.15 Equivalent-viscous damping due to plastic yielding for various values of deformation

$$\zeta_h = \frac{E_h}{4\pi E_s} \tag{3.7}$$

For a 10-cm deformation cycle, $E_h = 213$ kNm and $E_s = 43.1$ kNm. Therefore, $\zeta_h = 213/(4\pi \cdot 43.1) = 0.39$ (or 39% of critical). Hysteretic damping is similarly computed for cycles of other amplitudes. Figure 3.15 shows a plot of hysteretic damping for various values of deformation. The hysteretic damping is zero for deformations smaller than 3.37 cm because the MF has not yet yielded.

The damping curve of Fig. 3.15 needs some adjustments before it can be used in the nonlinear-static analysis. These adjustments are discussed next. During seismic response, the peak deformation occurs only once; rest of the times the deformation is less than the peak. Since the damping for smaller amplitude cycles is less, the damping is adjusted as follows. For $D = 10$ cm, the damping is 0.39 according to Fig. 3.15. The average damping for deformations between 0 and 10 cm is calculated by taking the area under the damping curve up to 10 cm and dividing that area by 10 cm (Fig. 3.16). This gives the average damping of $1.76/10 = 0.176$ (18% of critical). The average damping is similarly computed for other values of D. Finally, the damping is not allowed to drop below 5% of critical to account for additional sources of energy dissipation besides plastic yielding. Figure 3.17 shows the

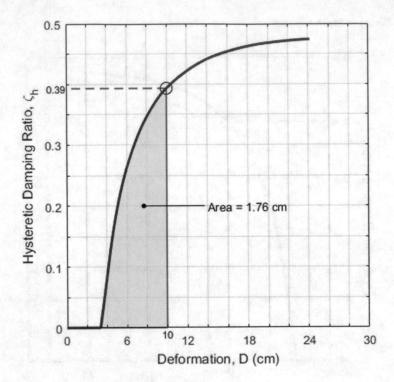

Fig. 3.16 Computing average hysteretic damping

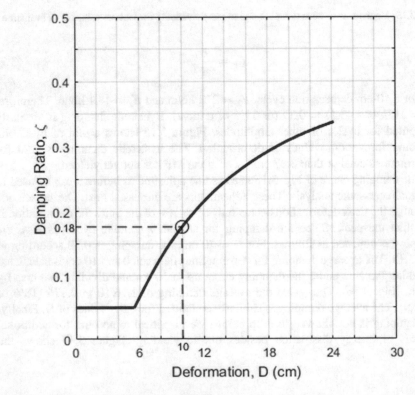

Fig. 3.17 Adjusted damping curve for the MF

adjusted damping curve for the system. This will be used to complete the nonlinear-static analysis.

3.7.4 Deformation-Versus-Damping Curve

In Fig. 3.18, the capacity curve of Fig. 3.12 is superimposed on the 500-year MRP response spectra (demand curves) for various values of damping. The response spectra (demand curves) in Fig. 3.18 are same as those in Fig. 3.2. The intersections of the capacity curve with the demand curves provide peak deformations for various assumed values of damping; these are shown in Fig. 3.19. In Fig. 3.18, the capacity curve stops short of the demand curve for 5% damping. This implies that the MF will collapse during the 500-year MRP ground motion if the damping of the MF were only 5% of critical. The plot in Fig. 3.19 is known as the deformation-versus-damping curve even though the deformations are shown along the horizontal axis. This curve represents the deformations of the MF for various assumed values of damping.

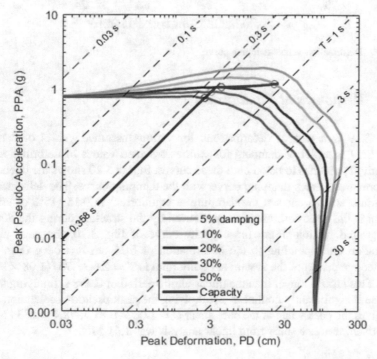

Fig. 3.18 Capacity curve superimposed on demand curves (response spectra) for 5%, 10%, 20%, 30%, and 50% damping

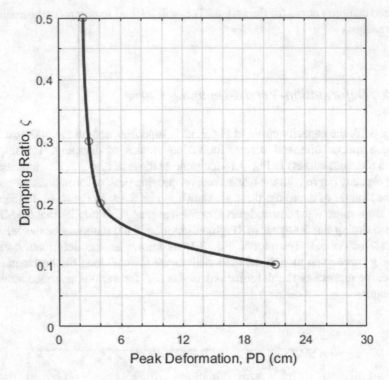

Fig. 3.19 Deformation-versus-damping curve

3.7.5 Responses at Equilibrium

Figure 3.19 is a plot of deformations for various assumed values of damping. Figure 3.17 is a plot of damping for various assumed values of deformation. The equilibrium point has to be on both these curves. Figure 3.20 shows the intersection of deformation-versus-damping curve with the damping curve. The deformation at equilibrium is 8.75 cm and the damping at equilibrium is 0.15 (15% of critical); 8.75 cm is the peak deformation experienced by the structure during the 500-year MRP ground motion. From the capacity curve of Fig. 3.10, the peak pseudo-acceleration corresponding to the deformation of 8.75 cm is 1.08 g (Fig. 3.21). The effective period of the system at equilibrium is $T = 2\pi\sqrt{8.75/(1.08 \times 981)} = 0.57$ s. This is 59% longer than the linear-elastic period of 0.358 s, implying that the response is significantly nonlinear. Multiplying the peak pseudo-acceleration by the mass gives the peak value of the base shear $Q = 1.08 \times 9.81 \times 78.4 = 833$ kN. As a comparison, the base shear from linear analysis was 1.41 MN.

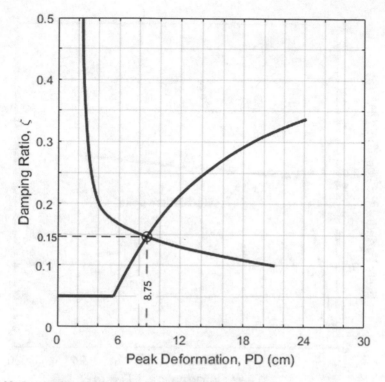

Fig. 3.20 Intersection of deformation-versus-damping curve (Fig. 3.19) with the damping curve (Fig. 3.17) to obtain damping and deformation at equilibrium

3.7.6 Expected Performance

Out of the total deformation of 8.75 cm, 3.37 cm deformation is elastic (Fig. 3.21) and the rest 5.38 cm is due to plastic yielding. From simple mechanics, the plastic hinge rotation is $\theta = 5.38/H = 0.0147$ radian. The limiting value of plastic rotation at which the columns collapse is 0.0568 radian (Table 3.1). The low-cycle fatigue damage is roughly proportional to the square of plastic deformation [11]. Since the limiting plastic rotation is 0.0568 radian (Table 3.1), the damage done by 0.0147 radian plastic rotation is $(0.0147/0.0568)^2 \times 100 = 7\%$. Therefore, the MF is practically undamaged by the 500-year MRP ground motion. It has ample fatigue life left to resist aftershocks. If nonstructural systems such as pipes and finishes can be shown to deform 8.75 cm without damage, the building can be considered to remain operational after the 500-year MRP ground motion.

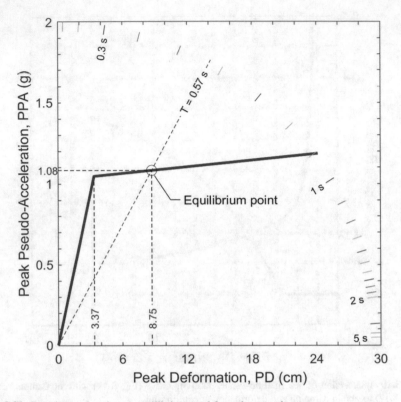

Fig. 3.21 Pseudo-acceleration at equilibrium read from the capacity curve

3.7.7 3D Visualization of Nonlinear-Static Analysis

It can be helpful to visualize the nonlinear-static analysis in three dimensions. The acceleration–deformation response spectra for various damping are stacked vertically to generate a demand surface. In Fig. 3.22, the demand surface is shown in colors ranging from blue to yellow. The capacity curve of Fig. 3.10 and the damping curve of Fig. 3.17 are combined to generate a 3D plot between deformation, pseudo-acceleration, and damping, known as the capacity-damping curve. It is the red curve in Fig. 3.22. The intersection of capacity-damping curve with the demand surface represents the equilibrium condition.

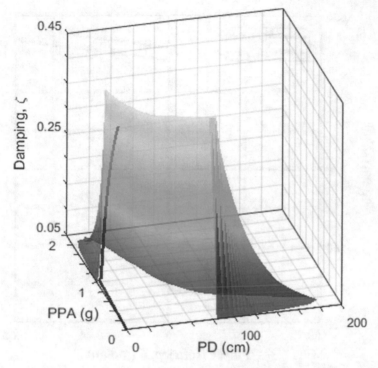

Fig. 3.22 Intersection of capacity-damping curve with the demand surface to obtain equilibrium

3.8 Nonlinear-Dynamic Analysis

The dynamic analysis of a nonlinear system is performed by solving the nonlinear equations of motion (EOM). Computer programs such as SAP 2000 [7] numerically solve the EOM in the background. However, the solution of nonlinear EOM can be easily corrupted by numerical errors. Therefore, a nonlinear-static analysis should always be carried out before a nonlinear-dynamic analysis. A computer model of the MF was generated in SAP 2000 [7]. Plastic hinges were assumed to form at the base of columns and the ends of beam. The moment–rotation relationships for plastic hinges were chosen to capture the net effect of strain hardening and P–Δ. According to Fig. 3.9, the ultimate strength of the MF increases 13% while the deformation D increases from 3.37 to 24.2 cm (or the plastic rotation increases from 0 to 0.0568 radian). Therefore, the moment capacity of the plastic hinges was assumed to increase 13% during a rotation increase from 0 to 0.0568 radian. Figure 3.23 shows the backbone stiffnesses of column and beam plastic hinges for dynamic analysis.

In Chap. 2, seven ground motion histories were generated to simulate the 500-year MRP ground motion at the SFBA site. Nonlinear-dynamic analysis was

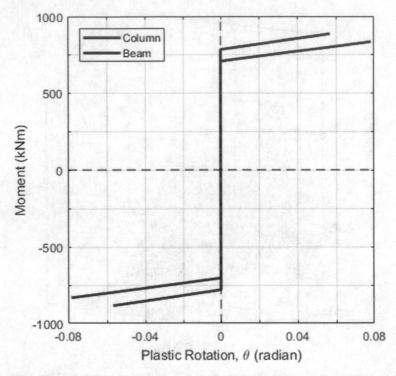

Fig. 3.23 Backbone stiffnesses of plastic hinges for dynamic analysis

performed seven times, for each of the simulated ground motions. Figure 3.24 shows the deformation responses to seven simulations of 500-year MRP ground motion. The peak deformation ranges from 4.67 cm (Simulation 6) to 5.87 cm (Simulation 4). The median value of peak deformations is 5.24 cm. Table 3.2 lists the peak responses of the MF to 500-year MRP ground motions. The median values in the last row of the table should be considered as the responses to the 500-year MRP ground motion. The base shear from nonlinear dynamic analysis is 836 kip. The pseudo-acceleration corresponding to base shear of 836 kip is 836/78.4/ $g = 1.09$ g.

A point corresponding to peak deformation of 5.87 cm and pseudo-acceleration of 1.09 g is drawn on the 500-year MRP ADRS in Fig. 3.25. The point is between 10% and 20% ADRS; it lies on the 18% damping ADRS. Therefore, the implicit damping in the nonlinear-dynamic analysis is 18% of critical. Based on Fig. 3.17, 18% damping seems too high for a deformation of 5.87 cm. It is difficult to know the source of high damping in the nonlinear-dynamic analysis without fully understanding the inner workings of the computer program used to perform nonlinear-dynamic analysis. It could be due to: (1) the manner in which various sources of damping are combined, (2) the shapes of hysteresis loops used in the analysis, or (3) numerically induced artificial damping to achieve stable solutions. The results of nonlinear-dynamic analysis should always be viewed with skepticism.

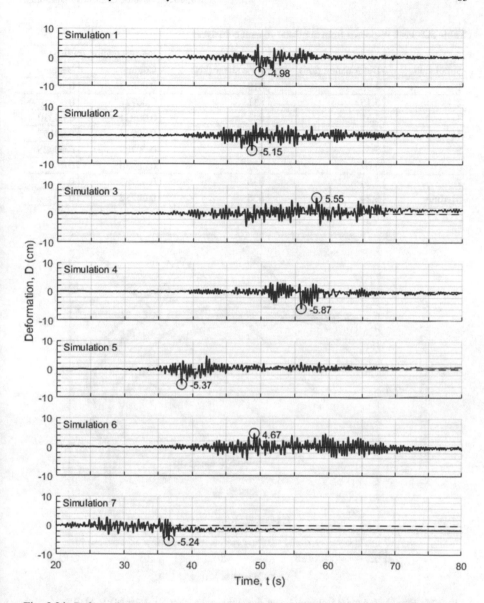

Fig. 3.24 Deformation responses to seven simulations of 500-year MRP ground motion at the SFBA site

Table 3.2 Peak responses from nonlinear-dynamic analyses

			Plastic rotation (radian)	
Simulation	Deformation (cm)	Base shear (kN)	Column	Beam
1	4.98	818	0.00673	0
2	5.15	831	0.00727	0
3	5.55	844	0.00834	0.00070
4	5.87	848	0.00917	0.00155
5	5.37	842	0.00789	0.00025
6	4.67	797	0.00580	0
7	5.24	836	0.00752	0
Median	**5.24**	**836**	**0.00752**	**0**

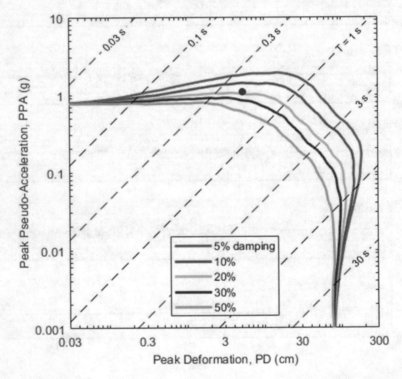

Fig. 3.25 Equilibrium point from nonlinear-dynamic analysis on 500-year MRP ADRS

3.9 Design of Connections

To ensure ductile performance of the MF, all connections should be strong enough to allow the structural elements to yield to their maximum extent. In Sect. 3.7.1 (subsection "strain hardening"), it was established that the ultimate moment capacities of columns and beam are 922 and 879 kNm, respectively. Therefore, the connections of columns to the foundation should be designed for >922 kNm moment, and the connections of beam to the columns should be designed for >879 kNm moment.

3.10   Uncertainty in Strength of Structural Elements

There is usually some uncertainty in the plastic moment capacity of beam and columns. If that is the case, lower and upper bound values of plastic moment capacity should be established based on material test reports or recommendations in ASCE 41 [6]. Lower bound estimates of plastic moment capacity should be used to calculate the deflections and plastic rotations. Upper bound estimates of plastic moment capacity should be used to calculate the base shear, overturning moment, and the connection moments.

3.11   Pros and Cons of Different Types of Analysis

Table 3.3 compares the results from different types of analysis. Linear analyses are applicable as long as the MF does not yield. In this particular case, the MF has clearly yielded; therefore, the results of linear analyses are not strictly applicable. For a nonlinear system, linear analyses overpredict the forces and underpredict the deformations. Building codes [12] prescribe R factors to reduce the forces obtained from the linear analyses. Although such analyses meet code requirements, they do not inform the true condition of the structure after the design ground motion. A nonlinear analysis is needed to assess the damage to the structure during the design ground motion.

Nonlinear analysis can be static or dynamic. A properly conducted, static analysis can predict the nonlinear response of a structure quite well. Static analysis is approximate but it is free of any hidden numerical errors. Dynamic analysis can predict the nonlinear responses more accurately if: (1) the analysis is not corrupted by numerical errors; (2) all sources of damping are adequately modeled; and (3) the ground motions used in the analysis match the response spectrum quite well. It is desirable to perform a static analysis before a dynamic analysis.

Table 3.3 Comparison of results from different methods of analysis

| Method of analysis | Deformation (cm) | Base shear (kN) | Plastic rotation (radian) | |
			Column	Beam
Linear static	5.85	1410	–	–
Linear dynamic	5.85	1410	–	–
Nonlinear static	8.75	833	0.0147	0.0147
Nonlinear dynamic	5.24	836	0.0075	0

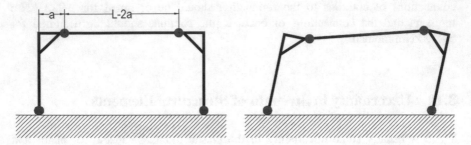

Fig. 3.26 Location of plastic hinges in a knee-braced frame

3.12   Knee-Braced Frame

So far, the plastic hinges in beam are assumed to form at the centerlines of columns. Due to finite width of columns, the beam plastic hinges are shifted to the faces of columns. This is also known as the rigid-end offset. If the beam is provided with haunches or knee braces, the plastic hinges are shifted farther from the column centerlines, as shown in Fig. 3.26. The shift in the plastic hinges changes the pushover curve for the frame. If L = horizontal distance between column centerlines, and a = distance of plastic hinges from column centerlines, the lateral force at yield is:

$$F_p = \frac{2\left(M_{pb} \times L/(L - 2a) + M_{pc}\right)}{H} \quad (3.8)$$

For $a = 0$, Eq. 3.8 reduces to Eq. 3.1. Eq. 3.5 is still valid except that θ is now the column plastic rotation $\theta_c = \theta$. The beam plastic rotation is as follows:

$$\theta_b = \theta_c \cdot \frac{L}{L - 2a} \quad (3.9)$$

For a knee-braced frame, the plastic-hinge rotation in the beam is greater than that in the columns. Since the limiting plastic rotation for the beam is also greater than the columns, a can be selected such that the beam and columns reach their limiting plastic rotations at the same deflection. This occurs when $L/(L - 2a) = 0.0783/0.0568 = 0.725$ or when $a = 1$ m. Figure 3.27 compares the capacity curve for an optimally designed knee-braced frame with the capacity curve for the MF. The knee-braced frame has higher strength but same deformability as the MF. The knee-braced frame has a seismic toughness of $ST = 2.83$ (m/s)^2 compared to $ST = 2.45$ (m/s)^2 for the MF.

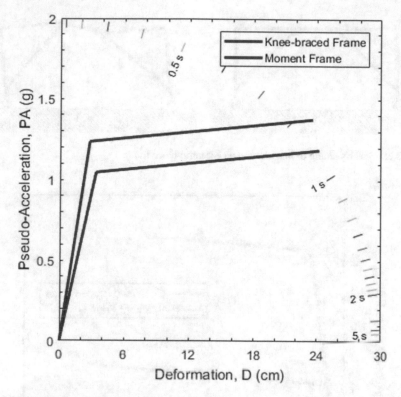

Fig. 3.27 Comparison between capacity curves of knee-braced frame and MF

3.13 Eccentric Braced Frame

The portion of beam between the two braces has a length of $L - 2a$ (Fig. 3.16). This is called the link beam. As a increases, the shear in the link beam increases. For a certain value of a, the link beam yields in shear rather than flexure. The shear yielding in steel beams is quite ductile as long as localized buckling can be prevented. An eccentric braced frame (EBF), shown in Fig. 3.28, utilizes shear yielding in the link beam to increase its deformability and damping. Figure 3.29 compares the capacity curve for an EBF with that for an MF. The EBF has seismic toughness of $ST = 3.53$ (m/s)^2 compared to $ST = 2.45$ (m/s)^2 for the MF.

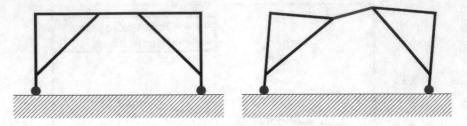

Fig. 3.28 Shear yielding of link beam in an eccentric brace frame

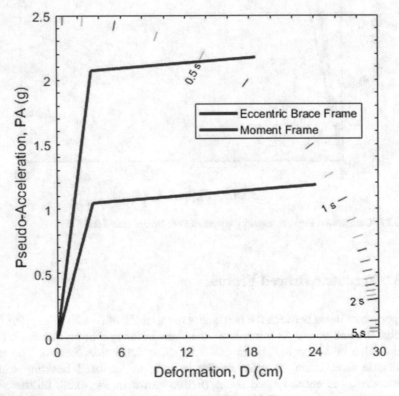

Fig. 3.29 Comparison between capacity curves for EBF and MF

3.14 Tension-Only Braced Frame

The brace elements in the knee-braced frame and the EBF were assumed to neither yield nor buckle; yielding was assumed to occur only in the beam and columns. Many times, the brace elements are designed for tension only; they buckle under compression (Fig. 3.30). Buckling of brace elements causes plastic hinges to form in

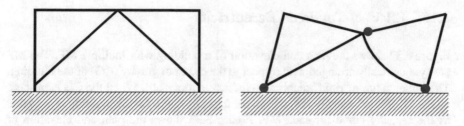

Fig. 3.30 Plastic hinges in a braced frame after buckling of compression member

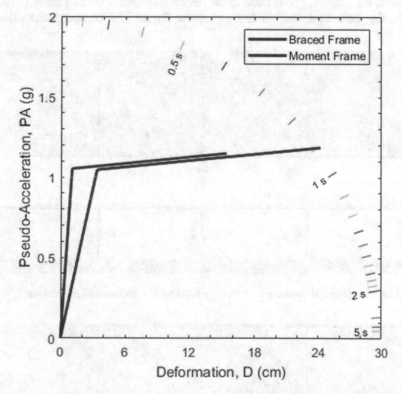

Fig. 3.31 Comparison between capacity curves for tension-only braced frame and MF

the beam and columns, as shown in Fig. 3.30. For the brace configuration shown in Fig. 3.30, the plastic rotation in the beam is twice the plastic rotation in the columns; therefore, the deformability of the system is controlled by the maximum allowable plastic rotation in the beam. Figure 3.31 compares the capacity curve for the tension-only braced frame with the capacity curve for the MF. The tension-only braced frame has seismic toughness of $ST = 1.6$ (m/s)^2 compared to $ST = 2.45$ (m/s)^2 for the MF.

3.15   Effect of Torsional Eccentricity

Figure 3.32 shows the plan and elevation of a building with multiple MF. The MF are symmetrically arranged with respect to the center of gravity (CG) of the building. The building has a "rigid" diaphragm (concrete roof slab). When the CG is pushed, all the MF deform equally. Therefore, the capacity curve for the building is the same as the capacity curve for each MF. Figure 3.33 shows the plan and elevation of another building for which the CG is shifted to the left due to the presence of a heavy rooftop equipment. When the CG is pushed, different MF move by different amounts; the leftmost MF (Frame 1) moves the maximum. When Frame 1 reaches its limiting deflection, the building collapses. Figure 3.34 compares the capacity

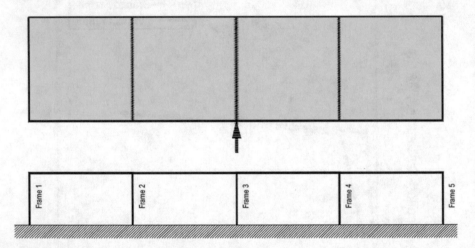

Fig. 3.32   Plan (top) and elevation (bottom) of a building with no torsional eccentricity

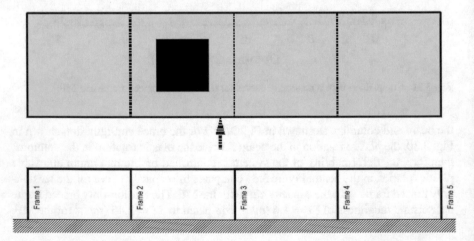

Fig. 3.33   Plan (top) and elevation (bottom) of a building with torsional eccentricity

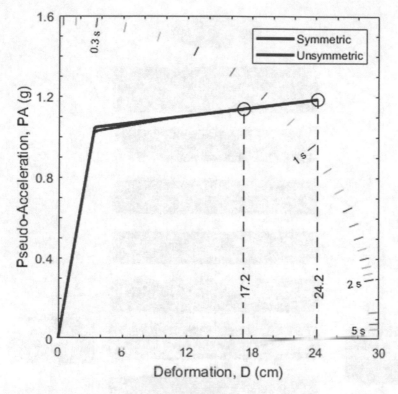

Fig. 3.34 Effect of torsional eccentricity on the capacity curve

curves for the symmetric and un-symmetric buildings. Torsional eccentricity reduces the deformability of the building from 24.2 cm to 17.2 cm and it reduces the seismic toughness ST of the building from 2.45 to 1.65 $(m/s)^2$. Torsional eccentricity always reduces the seismic performance of a building.

3.16 Seismic Design Steps

Seismic design based on linear analysis is arbitrary because structures other than nuclear power plants and major dams are not expected to remain linear during rare ground motions with MRP of about 2500 years. Even during 500-year MRP ground motions, limited amounts of nonlinearity can be tolerated as long as the desired performance goal is met. Nonlinear analysis can be static or dynamic. Nonlinear-dynamic analysis is not always practical because it can be corrupted by numerical errors. Therefore, the best choice of analysis for seismic design is a nonlinear-static analysis. Figure 3.35 presents a flowchart of seismic design steps based on nonlinear-static analysis.

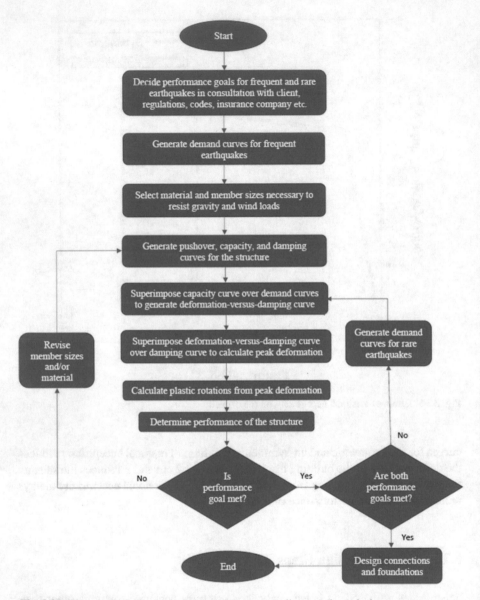

Fig. 3.35 Flowchart of seismic design steps based on nonlinear-static analysis

3.17 Summary

1. Plastic yielding in building frames can be a significant source of deformability and damping.
2. Plastic yielding does some damage to a building frame, but small amounts of plastic yielding can be accepted in operational performance and significant amounts of plastic yielding can be accepted in collapse prevention.

3. Connections in building frames should be strong enough to allow the structural members to yield to their maximum capacity. The design strength of connections should account for strain hardening of structural members.
4. There is some uncertainty in the strength of structural members. Lower estimates of strength should be used to calculate the deflections and plastic rotations. Upper estimates of strength should be used to calculate the base shear, overturning moment, and connection forces.
5. To achieve a ductile system, brittle failures should be avoided. Some sources of brittle failures are: (1) breakage of connections, (2) shear and axial failure of concrete elements, and (3) buckling of steel elements.
6. A well-engineered structure should be strong enough to avoid damage during frequent ground motions, and it should be ductile enough to avoid collapse during rare ground motions. The definitions of frequent and rare ground motions can vary from structure to structure, but usually frequent ground motion is exceeded with an MRP of 500 years and rare ground motion is exceeded with an MRP of 2500 years.
7. For a brittle structure, linear analysis is sufficient. For a ductile structure, nonlinear analysis is needed to assess the full seismic resistance of the structure.
8. Nonlinear analysis can be static or dynamic. Static analysis makes use of the response spectra of ground motion. Dynamic analysis makes use of the ground motion histories.
9. Nonlinear-dynamic analysis can be more accurate than the nonlinear-static analysis, if: (1) the ground motion histories used in the dynamic analysis match the site-specific response spectra well; (2) the solution of nonlinear EOM is not corrupted by numerical errors; and (3) all sources of damping are adequately considered.
10. The results of nonlinear-dynamic analysis should not be blindly trusted. Sanity checks, including the test of implicit damping, should be performed to assess the accuracy of nonlinear-dynamic analysis. A nonlinear-dynamic analysis should always be preceded with a nonlinear-static analysis.
11. Seismic toughness ST is a measure of structure's strength and deformability. It is an important indicator of structure's performance during earthquakes. Structures should be designed to maximize their seismic toughness and damping.
12. Optimally designed knee braces and eccentric braces can increase the seismic toughness of a building frame.
13. Torsional irregularity reduces the seismic toughness of a building. Therefore, it always reduces the seismic performance of a building.

References

1. Hamburger, R. O., Krawinkler, H., Malley, J. O., & Adan, S. A. (2009, June). Seismic design of steel special moment frames: A guide for practicing engineers. *NIST GCR 09-917-3*, U.S. Department of Commerce, National Institute of Standards and Technology.
2. AISC. (2016). Seismic provisions for structural steel buildings. *ANSI/AISC 341-16*. American Institute of Steel Construction, Inc., Chicago, IL.

3. AISC. (2016). Prequalified connections for special and intermediate steel moment frames for seismic applications including supplement no. 1. *ANSI/AISC 358-16*. American Institute of Steel Construction, Inc., Chicago, IL.
4. Engelhardt, M. D., Winneberger, T., Zekany, A. J., & Potyraj, T. J. (1996, August). The Dogbone connection: Part II. Modern Steel Construction.
5. Newell, J., & Uang, C.-M. (2006, December). Cyclic behavior of steel columns with combined high axial load and drift demand. *Report No. SSRP-06/22*, Department of Structural Engineering, University of California, San Diego La Jolla, CA.
6. ASCE. (2016). Seismic evaluation and retrofit of existing buildings. *ASCE 41-16*, American Society of Civil Engineers, Reston, VA.
7. CSI. (2011). *SAP 2000*. Berkeley, CA: Computers and Structures.
8. Paulay, T. (1978). A consideration of P-delta effects in ductile reinforced concrete frames. *Bulletin of the New Zealand National Society for Earthquake Engineering., 111*(3), 151–160.
9. Clough, R. W., & Penzien, J. (1976). *Dynamics of structures*. McGraw Hill Publications.
10. Chopra, A. K. (2017). *Dynamics of structures* (5th ed.). Pearson Publication.
11. Malhotra, P. K. (2002). Cyclic-demand spectrum. *Journal of Earthquake Engineering and Structural Dynamics, 31*(7), 1441–1457.
12. ASCE. (2016). Minimum design loads for building and other structures. *ASCE Standard No. ASCE 7-16*, American Society of Civil Engineers, Reston, VA.

Chapter 4
Seismic Response of Multistory Buildings

Nomenclature

ζ	Viscous damping ratio
2D	Two dimensional
ADRS	Acceleration–deformation response spectrum
CG	Center of gravity
EOM	Equation(s) of motion
g	$9.81 \ \text{m/s}^2$ = acceleration due to gravity
H	Height of CG
MF	Moment frame(s)
MRP	Mean return period
OTM	Overturning moment
PD	Peak deformation
PPA	Peak pseudo-acceleration
Q	Base shear
SDOF	Single degree of freedom system
SFBA	San Francisco Bay Area
SRSS	Square root of sum of squares
ST	Seismic toughness (area below the capacity curve)
T	Natural period of structure

4.1 Introduction

Figure 4.1 shows a five-story moment frame (MF) with rigid (fully restrained) beam-column moment connections. The MF could be part of a multistory building or some other industrial structure. Horizontal motion of the ground generates moments in beams and columns, causing them to bend and sway (Fig. 4.1). As discussed in

© Springer Nature Switzerland AG 2021
P. K. Malhotra, *Seismic Analysis of Structures and Equipment*,
https://doi.org/10.1007/978-3-030-57858-9_4

Fig. 4.1 Sketch of a five-
story building frame

Chap. 3, it is highly desirable that the connections be stronger than the connected
members so that plastic hinges can form in beams and columns before brittle failure
of connections. It is also desirable that the columns be stronger than the beams so
that a "soft story" is not created by yielding at both ends of columns in a story
[1, 2]. When created, a "soft story" tends to deform more than other stories, thus
increasing its chance of collapse by the P–Δ effect [1–3]. Beams are sometimes
intentionally weakened to ensure that yielding occurs in beams before failure of
connections or yielding of columns [2]. A properly designed MF experiences
flexural yielding in beams before any brittle failures. In steel MF, brittle failures
occur in connections. In concrete MF, brittle failures also include shear and axial
failures of structural members. Specific guidelines should be followed to ensure
ductile response of MF [1, 2, 4, 5].

4.2 Properties of MF

The MF, shown in Fig. 4.1, is fabricated from W24 × 162 columns in the lower three
stories, W21 × 147 columns in the upper two stories, W21 × 147 beams at the lower
three levels, and W18 × 97 beams at the upper two levels. The steel type is ASTM
A36 with an expected yield strength of 250 MPa. Table 4.1 lists the important
structural properties of the MF. The lumped mass at each level is 78.4×10^3 kg.
Therefore, the total lumped mass is $78.4 \times 10^3 \times 5 = 392 \times 10^3$ kg. It is assumed
that the MF is ductile, i.e., the connections are strong enough to allow the steel
members to yield to their maximum capacity.

Table 4.1 Properties of the MF shown in Fig. 4.1

Length of beam	7.32 m (each level)
Height of column	3.66 m (each story)
Lumped mass	78.4×10^3 kg (each level)
Moment of inertia of column cross section	215,530 cm^4 (first, second, and third story)
	151,330 cm^4 (fourth and fifth story)
Moment of inertia of beam cross section	151,330 cm^4 (first, second, and third level)
	72,955 cm^4 (fourth and fifth level)
Plastic moment capacity of column	1904 kNm (first, second, and third story)
	1517 kNm (fourth and fifth story)
Plastic moment capacity of beam	1517 kNm (first, second, and third level)
	858 kNm (fourth and fifth level)
Maximum allowable plastic hinge rotation in columns	0.0568 radian [6]
Maximum allowable plastic hinge rotation in beams	0.0783 radian [6]

4.3 Ground Motion

In this chapter, the response of the MF is computed during the 500-year MRP ground motion at a site in the San Francisco Bay Area (SFBA). The 500-year MRP ground motion for the SFBA site was determined in Chap. 2. Figure 4.2 shows the 500-year MRP response spectra for various values of damping. The response spectra are shown in the acceleration–deformation format. These are the demand curves for the site. The peak pseudo-acceleration *PPA* is read along the vertical axis and peak deformation *PD* is read along the horizontal axis. The natural period *T* is shown by parallel diagonal lines. The *PPA*, *PD*, and *T* are related to each other by the expression: $PD = PPA \times (T/2\pi)^2$. Figure 4.3 shows one of the seven spectrum-compatible ground motion history for 2D analyses; this was also generated in Chap. 2. Different types of analysis are discussed next.

4.4 Linear-Static Analysis

The MF has five lumped masses which can move relative to each other. Therefore, it is a multi-degree-of-freedom (MDOF) system. The basis of linear-static analysis is that the MDOF system can be represented by a set of linear, viscously damped single-degree-of-freedom (SDOF) systems, which can be analyzed by using the response spectrum of ground motion. The MF is linear before any yielding occurs in beams and columns. For a linear system, eigen-value analysis [7, 8] can be performed to generate a set of SDOF systems representing the various modes of vibration of the MDOF system. Each mode is characterized by its: (1) natural period,

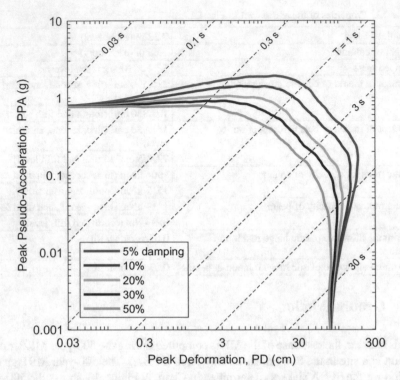

Fig. 4.2 500-year MRP ADRS for various values of damping for the SFBA site

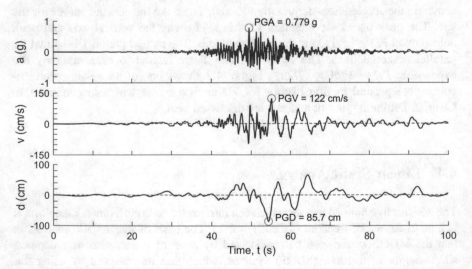

Fig. 4.3 One of seven 500-year MRP spectrum-compatible ground motion history

(2) mass, (3) height, (4) damping, and (5) normalized mode shape. The generation of modal properties is discussed next.

4.4.1   Modal Properties

A modal (eigen-value) analysis of the MF is performed by using SAP 2000 [9]. Figure 4.4 displays the mode shapes obtained from the modal analysis. Table 4.2 lists the mode-shape values. Each mode shape is associated with a natural period listed in the second row of Table 4.3. The natural frequency of each mode is reciprocal of its period; it is listed in the third row of Table 4.3. One way to understand the meaning of mode shapes is to imagine that the ground is shaken by purely sinusoidal motion of a specific frequency. If the frequency of the ground motion matches the natural frequency of a specific mode, then only that mode is excited, and the deformed shape of the structure matches the shape of that mode. For example, if the ground is shaken by 2.8 Hz frequency, then the deformed shape of the MF matches the second mode

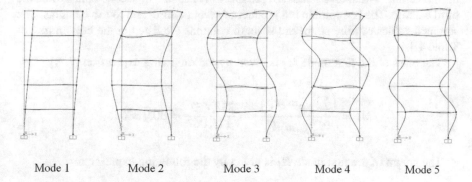

Mode 1   Mode 2   Mode 3   Mode 4   Mode 5

Fig. 4.4 Mode shapes of the MF shown in Fig. 4.1

Table 4.2 Mode shapes of the five-story MF shown in Fig. 4.1

Level, j	Mode, i	1	2	3	4	5
1	ϕ_{i1}	0.1334	−0.4069	0.7580	−0.8947	0.8222
2	ϕ_{i2}	0.3613	−0.8325	0.7089	0.1716	−0.9374
3	ϕ_{i3}	0.5954	−0.7838	−0.4429	0.7368	0.7256
4	ϕ_{i4}	0.8301	−0.0789	−0.8035	−0.8670	−0.3764
5	ϕ_{i5}	1.021	0.8689	0.5617	0.3314	0.1072

Table 4.3 Modal periods and frequencies

Mode, i	1	2	3	4	5
Natural period, T_i (s)	1.086	0.357	0.175	0.112	0.0825
Natural frequency, f_i (Hz)	0.921	2.8	5.71	8.93	12.12

Table 4.4 Mode 1 factors

Level, j	ϕ_{1j}	m_j (kg)	h_j (m)	$m_j\phi_{1j}$	$m_j\phi_{1j}^2$	$m_jh_j\phi_{1j}$
1	0.1334	78,400	3.66	10,454	1395	38,300
2	0.3613	78,400	7.32	28,307	10,227	207,200
3	0.5954	78,400	11	46,649	27,773	512,100
4	0.8301	78,400	14.6	65,042	53,992	952,000
5	1.021	78,400	18.3	80,000	81.681	1,463,000
			$\Sigma =$	230,452	175,067	3,173,100

shape displayed in Fig. 4.4. Real ground motions are composed of many different frequencies. Therefore, all modes are excited, but some modes are excited more than others, as will be seen later in this chapter.

Some additional properties (including mass and height of each mode) are determined next. The second column of Table 4.4 lists the shape of Mode 1 copied from Table 4.2. The third column lists the lumped masses at each level. The fourth column lists the height of each lumped mass. The fifth column is a product of second and third columns. The sixth column is a product of the square of second column and the third column. The last column is a product of second, third, and fourth columns. The summed values of fifth, sixth, and seventh columns are listed in the bottom row of Table 4.4.

The mass of the first mode M_1 is given by the following expression [7, 8]:

$$M_1 = \frac{\left(\sum_{j=1}^{5} m_j\phi_{1j}\right)^2}{\sum_{j=1}^{5} m_j\phi_{1j}^2} = \frac{230,452^2}{175,067} = 303,500 \text{ kg}$$

The height of the first mode H_1 is given by the following expression:

$$H_1 = \frac{\sum_{j=1}^{5} m_j h_j \phi_{1j}}{\sum_{j=1}^{5} m_j \phi_{1j}} = \frac{3,173,100}{230,452} = 13.8 \text{ m}$$

The normalizing factor for the first mode P_1 is given by the following expression:

$$P_1 = \frac{\sum_{j=1}^{5} m_j \phi_{1j}}{\sum_{j=1}^{5} m_j \phi_{1j}^2} = \frac{230,452}{175,067} = 1.32$$

P_1 is multiplied by the first mode shape, listed in the third column of Table 4.2, to obtain the normalized mode shape. The normalized mode shape relates the modal responses to the responses of the MF. The use of the normalized mode shapes will become clearer later. The modal properties were similarly computed for all five modes. Table 4.5 provides a complete list of modal properties. Table 4.6 provides the normalized mode shapes.

Table 4.5 Modal properties

Mode, i	1	2	3	4	5
Period, T_i (s)	1.086	0.357	0.175	0.112	0.0825
Mass, M_i (kg)	303,500 (77%)	53,500 (14%)	21,500 (5.5%)	9500 (2.5%)	4000 (1%)
Height, H_i (m)	13.8	1.17	2.06	1.06	1.66
Damping, ζ_i	0.05	0.05	0.05	0.05	0.05

Table 4.6 Normalized mode shapes of the moment frame shown in Fig. 4.1

Level, j	Mode 1, ϕ_{1j}	Mode 2, ϕ_{2j}	Mode 3, ϕ_{3j}	Mode 4, ϕ_{4j}	Mode 5, ϕ_{5j}
1	0.176	0.225	0.265	0.209	0.126
2	0.476	0.459	0.248	−0.040	−0.143
3	0.784	0.433	−0.155	−0.172	0.111
4	1.093	0.044	−0.281	0.202	−0.057
5	1.344	−0.480	0.197	−0.077	0.016

The first mode has the longest period of $T_1 = 1.086$ s. The mass of the first mode is $M_1 = 303,500$ kg which is 77% of the total mass. The masses of all modes add up to the total mass of the structure. The modal masses are used to calculate the horizontal force (base shear). The first modal height is $H_1 = 13.8$ m, which is 75% of the total height of the moment frame. The modal heights are used in the calculation of the overturning moment. As is typical, modal damping for linear analysis is assumed to be 5% of critical. The normalized mode shapes in Table 4.6 relate the modal deformations to the deflections of the MF. For example, 1 cm deformation of the first mode corresponds to 0.176 cm deflection at the first level and 1.34 cm deflection at the fifth level of the MF. Similarly, 1 cm deformation of the second mode corresponds to 0.225 cm deflection at the first level and −0.48 cm deflection at the fifth level of the MF. Note that positive modal deformation can result in negative deflections at some levels.

Figure 4.5 shows the five-story MF modeled by five SDOF systems. If the five SDOF systems are shaken by the same ground motion as the MF, then the total shear and overturning moment at the base of five SDOF systems will be same as those at the base of the MF. The modal masses and heights are drawn to scale to visualize the relative contributions of various modes. The CG of the building is at the same height as the CG of five modal masses.

4.4.2 Modal Responses

Table 4.7 lists the modal responses from the linear-static analysis. The second and third rows list the modal masses and heights from Table 4.5. Corresponding to the modal periods in the second row of Table 4.5, the peak pseudo-accelerations PPA_i are read from the 5% damping pseudo-acceleration response spectrum (Fig. 4.6);

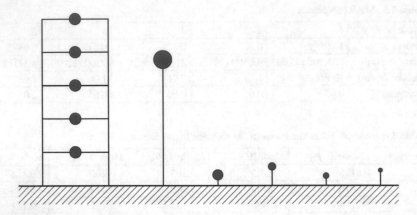

Fig. 4.5 Five-story MF modeled by 5 SDOF systems

Table 4.7 Peak modal responses

Mode, i	1	2	3	4	5	SRSS
Mass, M_i (kg)	303,500	53,500	21,500	9500	4000	–
Height, H_i (m)	13.8	1.17	2.06	1.06	1.66	–
PPA_i (g)	1.33	1.83	1.38	1.1	0.949	–
PD_i (cm)	39	5.78	1.05	0.34	0.16	–
$Q_i = M_i \cdot PPA_i$ (MN)	3.96	0.96	0.29	0.103	0.038	**4.1**
$OTM_i = Q_i \cdot H_i$ (kNm)	54.5	1.12	0.6	0.11	0.063	**55**
$D_{i5} = \phi_{i5} \cdot PD_i$ (cm)	52.4	2.8	0.21	0.03	0.02	**52.4**

they are listed in the fourth row of Table 4.7. The modal deformations are obtained from the expression: $PD_i = PPA_i \times (T_i/2\pi)^2$; they are listed in the fifth row of Table 4.7. Alternatively, modal accelerations and deformations can be simultaneously read from the 5% damping ADRS (Fig. 4.7). Modal base shears Q_i are obtained by multiplying the modal accelerations with the modal masses; they are listed in the sixth row of Table 4.7. Modal overturning moments OTM_i are obtained by multiplying the modal base shears with the modal heights; they are listed in the seventh row of Table 4.7. The deflections at various levels of the MF are obtained by multiplying modal deformations with the normalized mode shapes (Table 4.6); the modal deflections at the fifth level D_{i5} are listed in the last row of Table 4.7.

The peak modal responses in various columns of Table 4.7 do not occur at the same time since modes have different natural periods. In a static analysis, the modal responses are combined by the SRSS (square-root-of-sum-of-squares) or other similar methods. The combined responses are listed in the last column of Table 4.7. This concludes the linear-static analysis.

The first mode base shear of 3.96 MN is 97% of the combined base shear of 4.1 MN. The first mode overturning moment of 54.5 MNm is 99% of the combined overturning moment of 55 MNm. The first mode deflection at the fifth level is

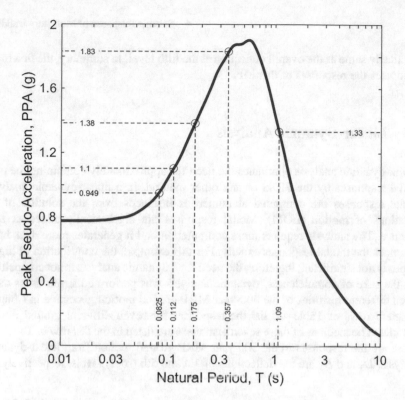

Fig. 4.6 Peak pseudo-accelerations for various modes of the MF

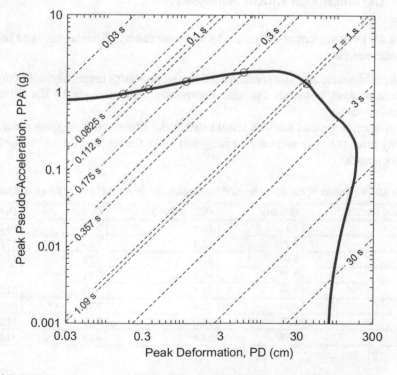

Fig. 4.7 Modal accelerations and deformations from the 5% damping ADRS

practically same as the overall deflection at the fifth level. In summary, the first mode dominates the responses of this MF.

4.5   Linear-Dynamic Analysis

Linear-dynamic analysis eliminates the need for approximately combining the peak modal responses by the SRSS or any other method. In a linear-dynamic analysis, modal responses are computed at numerous instances from the solution of the equations of motion (EOM). Modal responses are algebraically added at each instance. The analysis requires more computation and it generates more data, but it combines the modal responses exactly. For this example, the contribution of higher modes is not significant, therefore, the need for a dynamic analysis is not compelling. For the sake of completeness, dynamic analysis was performed seven times using seven different histories of the 500-year MRP ground motion generated in Chap. 2. Different rows of Table 4.8 list the responses for seven different ground motion histories. The medians of these seven responses are listed in the last row of Table 4.8. These are the responses to the 500-year MRP ground motion from linear-dynamic analysis. D_1 and D_5 are the deflections of 1st and 5th (roof) levels, respectively.

4.6   Comments on Linear Analyses

There are three sources of difference between the results of linear-static and linear-dynamic analyses:

1. Linear-dynamic analysis combines the modal responses accurately, while linear-static analysis combines the modal responses approximately by the SRSS or similar methods.
2. Site-specific ground motions should match the site-specific response spectrum. Any discrepancy in spectral matching will make the results of dynamic analysis less accurate.

Table 4.8 Responses of the MF to seven different histories of the 500-year MRP ground motion

Ground motion	Q (MN)	OTM (MNm)	D_1 (cm)	D_5 (cm)
1	4.14	55.5	7.16	53.2
2	4.22	56.4	7.33	53.5
3	3.96	52	6.75	49.2
4	3.77	50.5	6.41	48.7
5	3.66	52.8	6.43	51.3
6	4.14	56.9	7.18	54.6
7	3.78	52.6	6.47	51.8
Median	**3.96**	**52.8**	**6.75**	**51.8**

3. Results of dynamic analysis can be corrupted by numerical errors. However, linear analyses are less prone to numerical errors than nonlinear analyses discussed later.

The results of linear-dynamic analysis are within 5% of those of linear-static analysis. For all practical purposes, the results of both analyses are identical in this case. The results of linear analyses are valid as long as no yielding occurs in the MF. Based on linear analyses, the maximum moment at the base of columns is 5.87 MNm; this is more than three times the plastic moment capacity of columns. The maximum moment in the lower story beams is 5.63 MNm; this is nearly four times the plastic moment capacity of beams. There are two options for the designer: (1) increase the capacities of columns and beams to exceed the demands imposed by the 500-year MRP ground motion or (2) allow beams and columns to yield and demonstrate that yielding does not preclude the MF from meeting its seismic performance objective. The first option relies purely on strength; it can result in a very expensive design. The second option relies on the deformability and damping of the MF to reduce the strength demand. However, a nonlinear analysis is needed to demonstrate that yielding is below the acceptable level. Two types of nonlinear analyses are discussed next.

4.7 Nonlinear-Static Analysis

The basis of nonlinear-static analysis is to generate an equivalent-viscous nonlinear SDOF model of the MF and to calculate its response from the response spectra of ground motion. As discussed in Chap. 3, nonlinear-static analysis requires the generation of pushover and damping curves. The pushover curve defines the strength and deformability of the structure; the damping curve defines its ability to dissipate energy.

4.7.1 Pushover Curve

The pushover curve is a plot between the lateral force and the lateral deformation of the system. The lateral force does not act at a single point; it acts on, and is proportional to, lumped masses at various levels of the MF. The centroid of the total lateral force F is at the center of gravity (CG) of the system, as shown in Fig. 4.8. The lateral deformation D is also measured at the CG of the system. As the MF is gradually pushed in the horizontal direction, the moments increase in the beams and columns. At some point, plastic hinges form at the base of columns. At some later points, plastic hinges also form at the ends of beams at various levels. Figure 4.8 shows the forces and moments applied on the columns after all plastic

Fig. 4.8 Forces and
moment acting on the
columns of the MF after the
formation of all plastic
hinges

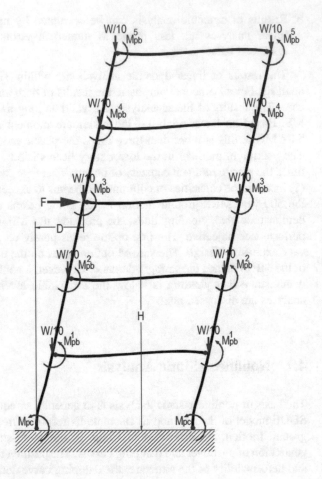

hinges have formed. Disregarding the effect of gravity, the lateral force after the
formation of all plastic hinges is given by the following expression:

$$F_p = \frac{2\left(\sum_j M_{pb}^j + M_{pc}\right)}{H}$$ (4.1)

in which M_{pb}^j = plastic moment capacity of the beam cross section at the jth level;
M_{pc} = plastic moment capacity of the column cross section at the base; and H =
height of the CG. Upon substituting $M_{pb}^1 = M_{pb}^2 = M_{pb}^3 = 1517$ kNm, $M_{pb}^4 = M_{pb}^5 =$
858 kNm, $M_{pc} = 1904$ kNm, and $H = 3 \times 3.66 = 10.98$ m from Table 4.1, $F_p =$
1490 kN.

The elastic stiffness, before the formation of any plastic hinges, is given by the
following expression:

$$k = M\left(\frac{2\pi}{T_1}\right)^2$$ (4.2)

Substituting $T_1 = 1.09$ s = first mode period of the MF and $M = 392 \times 10^3$ kg = total lumped mass, Eq. 4.2 gives $k = 13.1$ MN/m. The maximum elastic deformation is:

$$D_e = \frac{F_p}{k} \tag{4.3}$$

Substituting $F_p = 1490$ kN and $k = 13.1$ MN/m, Eq. 4.3 gives $D_e = 11.3$ cm. The elastic portion of the pushover curve ($D \leq D_e$) is generated by using the following relationship

$$F = k \cdot D \tag{4.4}$$

For deformation $D \leq D_e$, the force increases linearly with D, but the plastic rotation θ remains zero. For $D > D_e$, the force remains fixed at $F = F_p$, but the plastic rotation θ increases with deformation. θ is the same for all beams and columns. The relationship between D and θ is as follows:

$$D = D_e + \theta \cdot H \tag{4.5}$$

The maximum allowable plastic hinge rotation is 0.0568 radian for columns and 0.0783 radian for beams (Table 4.1). Since columns and beams experience the same plastic rotation, the allowable rotation in columns controls the deflection at collapse, i.e., $11.3 + 0.0568 \times H = 73.7$ cm. Figure 4.9 shows the simplified pushover curve for the MF from Eqs. 4.1–4.5. In a more accurate pushover curve, the transition from elastic to plastic is more gradual as plastic hinges in columns and beams do not appear at the same time.

The pushover curve of Fig. 4.9 ignores any increase in plastic moment with rotation due to strain hardening. The pushover curve also ignores the effect of gravity, aka the P–Δ effect [3]. The pushover curve is refined as follows:

Strain hardening. ASCE 41 [6] recommends that the post-yield stiffness be assumed equal to 3% of the elastic stiffness. This is arbitrary because strain hardening should not depend on the elastic stiffness. For lack of better information, ASCE 7-41 [6] recommendation is followed. Figure 4.10 shows the pushover curve with strain hardening. Strain hardening causes the ultimate lateral strength of the MF to increase from 1490 to 1730 kN. This is a 16% increase in lateral strength, while the plastic rotation increases from 0 to 0.0568 radian. Therefore, the ultimate moment capacity of columns is $1904 \times 1.16 = 2209$ kNm. The allowable plastic rotation in beams is higher; therefore, the increase in moment capacity in beams is $16/0.0568 \times 0.0783 = 22\%$. The ultimate moment capacity of beams at first three levels is $1517 \times 1.22 = 1850$ kNm and at upper two levels is $858 \times 1.22 = 1050$ kNm.

P–Δ effect. Referring to Fig. 4.8, note that the weight of the structure applies an additional moment which is in the same direction as the moment applied by the lateral force F. Both moments are resisted by the MF. The weight of the structure

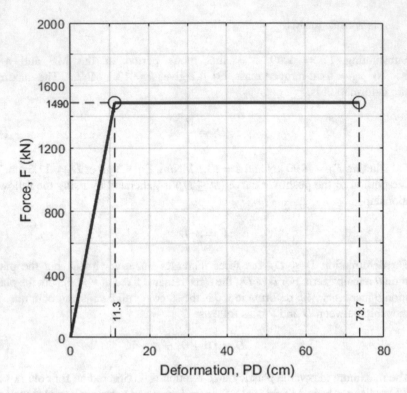

Fig. 4.9 Simplified pushover curve for the MF shown in Fig. 4.8

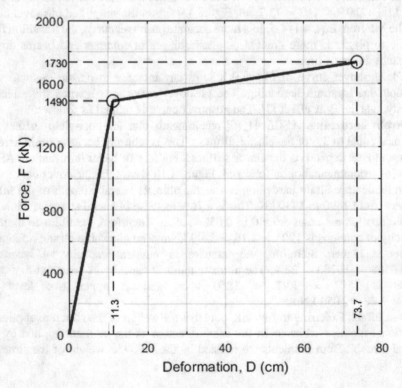

Fig. 4.10 Pushover curve with strain hardening

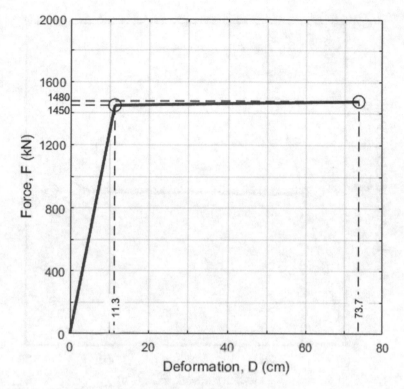

Fig. 4.11 Pushover curve with strain hardening and P–Δ effect

reduces the lateral force F for a given deflection D. P–Δ effect is considered as follows: (1) for a given value of D, F is read from Fig. 4.10 and (2) F is reduced by an amount DW/H, where W = the weight of the structure and H = the height of the CG. Figure 4.11 shows the refined pushover curve after considering strain hardening and P–Δ effect. Due to P–Δ, the yield strength of the MF reduces from 1490 to 1450 kN, and the ultimate strength of the MF reduces from 1730 to 1480 kN. The ultimate strength of 1480 kN is only 2.1% higher than the yield strength of 1450 kN. This increase occurs while the deformation D increases from 11.3 to 73.7 cm or the plastic rotation increases from 0 to 0.0568 radian. Therefore, the increase in strength is 2.1/0.0568 = 36% per radian.

The P–Δ effect is much more pronounced for the five-story MF than for the one-story MF discussed in Chap. 3. This is because the 5-story MF deforms much more than the one-story MF.

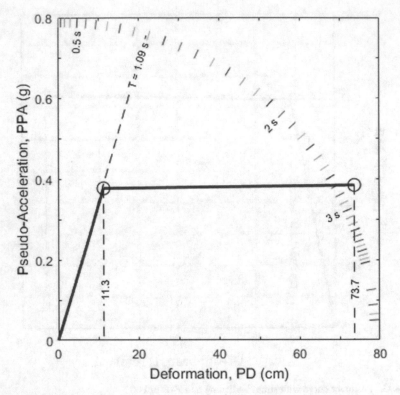

Fig. 4.12 Capacity curve for the MF shown in Fig. 4.1

4.7.2 Capacity Curve

The pushover curve of Fig. 4.11 is converted into the capacity curve by dividing the force along the vertical axis by the total lumped mass M. Figure 4.12 shows a plot of the capacity curve. The vertical axis in Fig. 4.12 is the pseudo-acceleration PA. Recall from Chap. 1 that the natural period depends on the ratio between the deformation and the pseudo-acceleration, i.e.,

$$T = 2\pi\sqrt{\frac{D}{PA}} \qquad (4.6)$$

With the help of Eq. 4.6, radial tick marks corresponding various periods are drawn in Fig. 4.12. Note that the period of the MF remains fixed at 1.09 s up to the formation of plastic hinges at 11.3 cm deformation. After that, the period increases with increase in deformation. The capacity curve ends at 73.7 cm deformation when the plastic rotation in columns reaches its limiting value of 0.0568 radian. The area below the capacity curve (Fig. 4.13) is called the seismic toughness ST. In general,

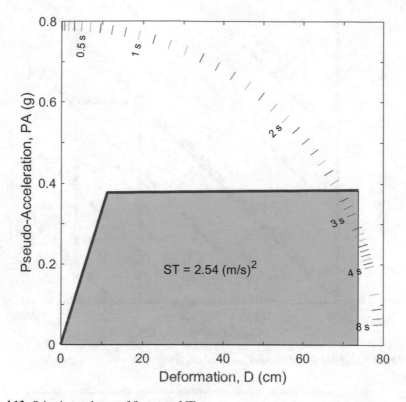

Fig. 4.13 Seismic toughness of five-story MF

tougher structures perform better during earthquakes. For the MF discussed here, $ST = 2.54$ (m/s)^2. ST has the units of energy/mass.

The capacity curve of Fig. 4.12 is redrawn on a logarithmic scale in Fig. 4.14. The radial tick marks in Fig. 4.12 are replaced by parallel diagonal lines in Fig. 4.14. Both linear and logarithmic plots of the capacity curve (Figs. 4.12 and 4.14) will be used in the analysis.

4.7.3 Damping Curve

During earthquakes, the MF deforms back and forth in a cyclic manner. Figure 4.15 shows the cyclic force–deformation relationship for the MF subjected to deformation cycles of various amplitudes. The force–deformation relationship is hysteretic— meaning that some energy is lost during each complete cycle. The energy loss occurs due to plastic hinge rotations and it equals the area enclosed within the force– deformation loop. This is the source of damping for the MF after yielding has occurred. The damping for various values of deformation is determined next.

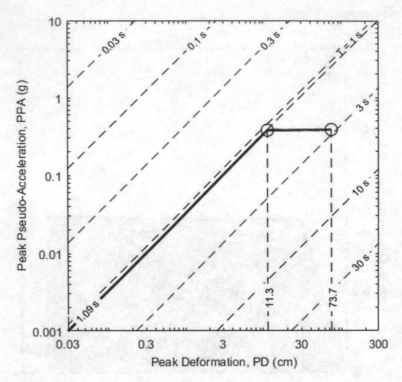

Fig. 4.14 Logarithmic plot of capacity curve shown in Fig. 4.12

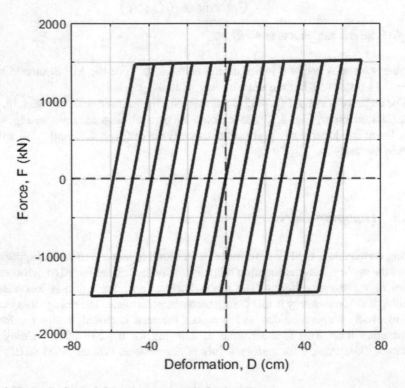

Fig. 4.15 Cyclic force–deformation relationship for the MF

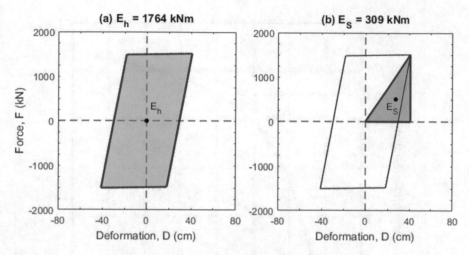

Fig. 4.16 Hysteretic and "strain" energies for 40-cm amplitude cycle

Figure 4.16a shows the force–deformation relationship for a hypothetical cycle of 40-cm amplitude. The energy loss E_h is the shaded area enclosed within the hysteresis loop. The "strain" energy E_s is one-half the product of the peak force and peak deformation; it is the area of the shaded triangle in Fig. 4.16b. The equivalent-viscous hysteretic damping is given by the following expression [8]:

$$\zeta_h = \frac{E_h}{4\pi E_s} \tag{4.7}$$

For a 40-cm deformation cycle, $E_h = 1764$ kNm and $E_s = 309$ kNm. Therefore, $\zeta_h = 1764/(4\pi \cdot 309) = 0.45$ (or 45% of critical). Hysteretic damping is similarly computed for cycles of other amplitudes. Figure 4.17 shows a plot of the hysteretic damping for various values of deformation. The hysteretic damping is zero for deformations smaller than 11.3 cm because the MF has not yet yielded.

The damping curve of Fig. 4.17 needs some adjustments before it can be used in nonlinear-static analysis. The purpose of nonlinear-static analysis is to predict the peak deformation. During seismic response, the peak deformation occurs only once; rest of the times the deformation is less than the peak. Since the damping for smaller amplitude cycles is less, the damping is adjusted as follows. For $D = 40$ cm, the damping is 0.45 according to Fig. 4.17. The average (adjusted) damping for deformations between 0 and 40 cm is calculated by taking the area under the damping curve up to 40 cm and dividing that area by 40 cm (Fig. 4.18). This gives the average (adjusted) damping of 9.04/40 = 0.23 (23% of critical). The adjusted damping is similarly computed for other values of D. Finally, the damping is not allowed to drop below 5% of critical because there are some other sources of energy dissipation besides plastic yielding. Figure 4.19 shows the adjusted damping curve for the MF; this will be used to complete the nonlinear-static analysis.

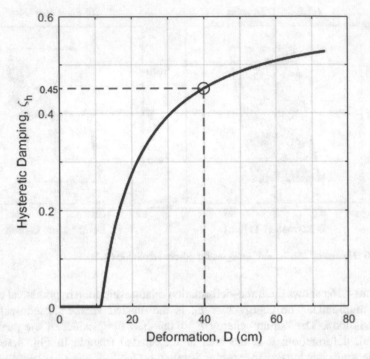

Fig. 4.17 Equivalent-viscous hysteretic damping for various values of deformation

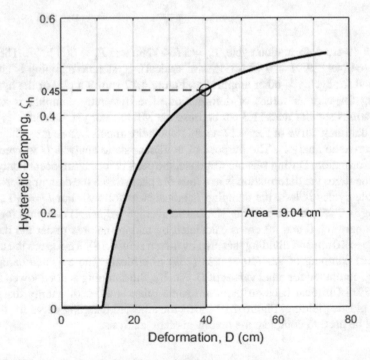

Fig. 4.18 Adjustment of equivalent-viscous damping

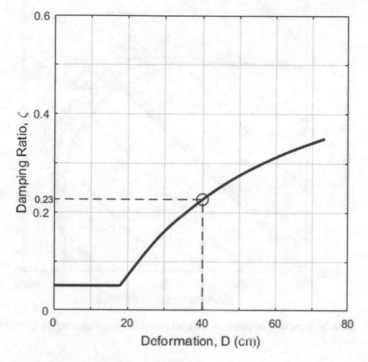

Fig. 4.19 Adjusted damping curve for the MF

4.7.4 Deformation-Versus-Damping Curve

In Fig. 4.20, the capacity curve of Fig. 4.14 is superimposed on the 500-year MRP response spectra (demand curves) of Fig. 4.2. The intersections of the capacity curve with the demand curves provide peak deformations for various assumed values of damping; they are shown in Fig. 4.21. In Fig. 4.20, the capacity curve stops short of the 5% damping demand curve. This implies that the MF would collapse if its damping were only 5% of critical. The plot in Fig. 4.21 is known as the deformation-versus-damping curve even though the deformations are shown along the horizontal axis. This curve represents the deformations of the MF for various assumed values of damping.

4.7.5 Responses at Equilibrium

The intersection of the deformation-versus-damping curve of Fig. 4.21 with the adjusted damping curve of Fig. 4.19 provides damping and deformation at equilibrium (Fig. 4.22). The deformation at equilibrium is 36 cm and the damping at equilibrium is 0.21 (21% of critical); 36 cm is the deflection of the CG (Fig. 4.8).

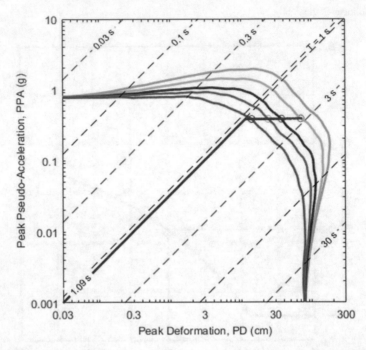

Fig. 4.20 Capacity curve superimposed on demand curves (response spectra) for 5, 10, 20, 30, and 50% damping

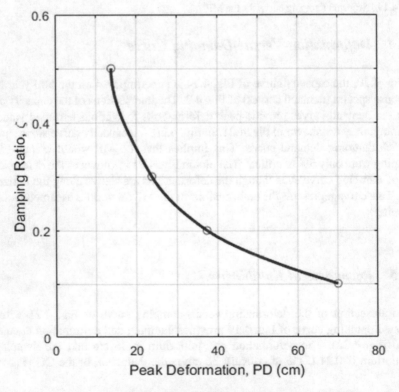

Fig. 4.21 Deformation-versus-damping curve for MF's response to 500-year MRP ground motion

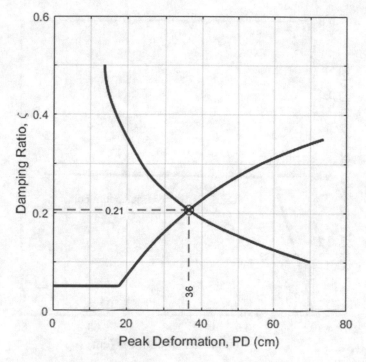

Fig. 4.22 Intersection of deformation-versus-damping curve (Fig. 4.21) with the adjusted damping curve (Fig. 4.19) to obtain damping and deformation at equilibrium

The roof deflection is $36/H \times h_5 = 55$ cm, where $H =$ height of CG and $h_5 =$ the roof height.

From the capacity curve of Fig. 4.12, the peak pseudo-acceleration corresponding to 36 cm deformation is 0.38 g (Fig. 4.23). The effective period of the system at equilibrium is $T = 2\pi\sqrt{36/(0.38 \times 981)} = 1.97$ s. This is 80% longer than the linear-elastic period of 1.09 s, implying that the response is significantly nonlinear. Multiplying the peak pseudo-acceleration by the mass gives the peak value of base shear $Q = 0.38 \times 9.81 \times 392 \times 10^3 = 1.46$ MN. The overturning moment is $OM = Q \times H = 1.46 \times H = 16$ MNm.

4.7.6 Expected Performance

Out of the total deformation of 36 cm, 11 cm deformation is elastic (Fig. 4.23) and 25 cm is due to plastic yielding. From simple geometry, the plastic-hinge rotation is $25/H = 25/1100 = 0.023$ radian. The low-cycle fatigue damage is roughly proportional to the square of the plastic rotation [10]. Since the limiting plastic rotation is 0.0568 radian (Table 4.1), the damage done by 0.023 radian is (0.023/

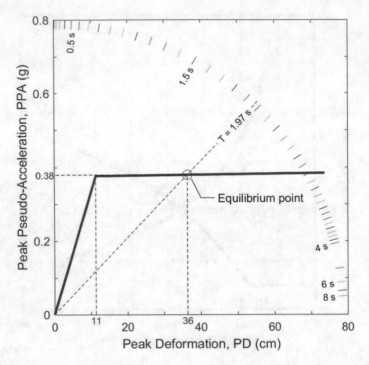

Fig. 4.23 Pseudo-acceleration at equilibrium read from the capacity curve

$0.0568)^2 \times 100 = 16\%$. Therefore, the MF is likely to need minimal repairs following the earthquake. In order for the building to remain operational following the 500-year MRP ground shaking, the MF will require slightly larger column and beam sections to bring the plastic rotation to <0.01 radian. In addition, nonstructural systems such as vertical pipe runs and wall finishes should be designed to absorb expected deflections without damage for the building to remain operational following the 500-year MRP ground motion.

4.7.7   3D Visualization of Nonlinear-Static Analysis

It can be helpful to visualize the nonlinear-static analysis in three dimensions. The acceleration–deformation response spectra for various damping are stacked vertically to generate a demand surface. In Fig. 4.24, the demand surface is shown in colors ranging from blue to yellow. The capacity curve of Fig. 4.12 and the damping curve of Fig. 4.19 are combined to generate a 3D plot between deformation, pseudo-acceleration, and damping, called the capacity–damping curve. It is the red curve in Fig. 4.24. The intersection of capacity–damping curve with the demand surface represents the equilibrium condition.

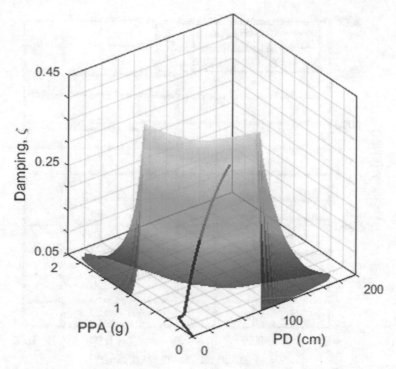

Fig. 4.24 Intersection of capacity–damping curve (red) with the demand surface (blue to yellow) to obtain equilibrium

4.8 Nonlinear-Dynamic Analysis

The dynamic analysis of a nonlinear system is performed by solving a set of nonlinear equations of motion (EOM). Computer programs such as SAP 2000 [9] numerically solve the EOM in the background. However, the solution of nonlinear EOM can be easily corrupted by numerical errors. Therefore, a nonlinear-static analysis should always be carried out before a nonlinear-dynamic analysis. A computer model of the MF was generated in SAP 2000 [9]. Plastic hinges were assumed to form at the base of columns and the ends of beams. According to Fig. 4.11, the ultimate strength of the MF increases 2.07%, while the deformation D increases from 11.3 to 73.7 cm (or the plastic rotation increases from 0 to 0.0568 radian). This increase is the net result of strain hardening and P–Δ effect. Therefore, the moment capacity of the plastic hinges was assumed to increase 2.07% during a rotation increase from 0 to 0.0568 radian. Figure 4.25 shows the backbone stiffnesses of column and beam plastic hinges for dynamic analysis.

The dynamic analysis was performed seven times using seven simulations of the 500-year MRP ground motion for the SFBA site. Figure 4.26 shows the histories of roof deflections D_5 during seven simulations of 500-year MRP ground motion. The

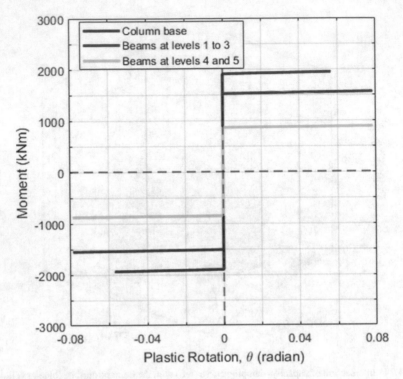

Fig. 4.25 Backbone stiffnesses of plastic hinges for dynamic analysis

peak roof deflections D_5 ranges from 33.4 cm (Simulation 4) to 67.5 cm (Simulation 3). The median value of roof deflection is 44.7 cm. Table 4.9 presents the results of the seven analyses. The median values in the last row of Table 4.9 are the responses to the 500-year MRP ground motion from dynamic analysis.

The deflection of the CG from nonlinear-dynamic analysis is 28 cm. The base shear from nonlinear-dynamic analysis is 1.49 MN; this corresponds to a pseudo-acceleration of 1490/392/g = 0.388 g. As a sanity check, the equilibrium point corresponding to peak deformation of 28 cm and peak pseudo-acceleration of 0.388 g is shown by a red dot on the 500-year MRP ADRS in Fig. 4.27. This point corresponds to a damping of 26% of critical. According to Fig. 4.19, 26% damping for 28 cm deformation seems too high. It is difficult to know the cause of higher than expected damping in the nonlinear-dynamic analysis without fully understanding the inner workings of the computer program used to perform nonlinear-dynamic analysis. It could be due to: (1) the manner in which various sources of damping are combined, (2) the shapes of hysteresis loops used in the analysis, or (3) numerically induced artificial damping to achieve stable solutions.

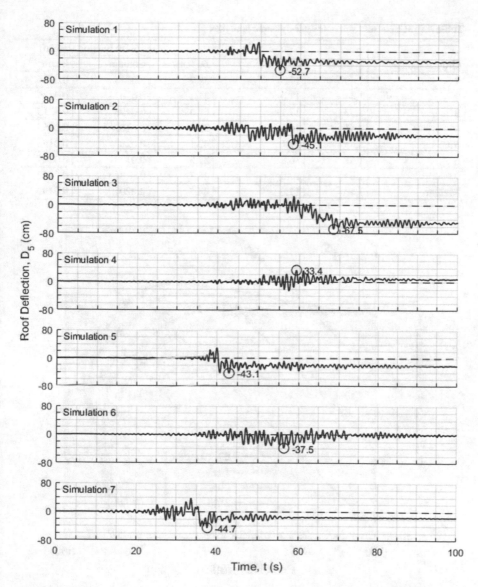

Fig. 4.26 Histories of roof deflection during seven simulations of 500-year MRP ground motion at the SFBA site

Table 4.9 Responses from nonlinear-dynamic analyses

Ground motion	Q (MN)	OTM (MNm)	D_5 (cm)	Plastic rotation (radian)	
				Column	Beam
1	1.49	17.5	52.7	0.0216	0.0246
2	1.44	17.0	45.1	0.0201	0.0195
3	1.57	17.1	67.5	0.0296	0.0315
4	1.41	16.6	33.4	0.0165	0.0153
5	1.43	17.1	43.1	0.0208	0.0196
6	1.53	16.8	37.5	0.0128	0.0149
7	1.64	17.0	44.7	0.0181	0.0192
Median	**1.49**	**17.0**	**44.7**	**0.0201**	**0.0195**

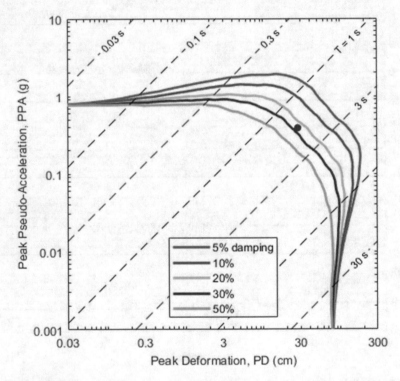

Fig. 4.27 Equilibrium point from nonlinear-dynamic analysis on 500-year MRP ADRS

4.9   Pros and Cons of Different Methods of Analysis

Linear-static and linear-dynamic analyses are applicable as long as the MF does not yield. Linear-static analysis is generally more conservative because it combines the modal contributions approximately by using the SRSS or similar rules. Linear-dynamic analysis is more accurate only if the ground motions used in analysis match the site-specific response spectrum quite well.

Table 4.10 Comparison of results from different methods of analysis

Method of analysis	Q (MN)	OTM (MNm)	D_5 (cm)	Plastic rotation (radian)	
				Column	Beam
Linear static	4.09	54.5	52.4	–	–
Linear dynamic	3.96	52.8	51.8	–	–
Nonlinear static	1.46	16	55	0.023	0.023
Nonlinear dynamic	1.49	17	45	0.020	0.020

Most structures continue to perform well into the nonlinear range. A nonlinear analysis is needed to assess the extent of plastic yielding hence damage to the structure. Nonlinear-static analysis is approximate but it is free of numerical errors and provides greater insight into the behavior of a structure.

Dynamic analysis can predict the nonlinear responses more accurately if: (1) the analysis is not corrupted by numerical errors; (2) all sources of damping are adequately modeled; and (3) the ground motions used in the analysis match the response spectrum quite well. It is desirable to perform a static analysis before a dynamic analysis.

Table 4.10 compares the responses from four different methods of analysis. Static and dynamic analyses provide similar results. Linear analyses overpredict forces and underpredict deformations. Linear analyses cannot predict plastic rotations.

4.10   Design of Connections

To ensure ductile performance of the MF, all connections should be strong enough to allow the structural elements to yield to their maximum extent. In Sect. 4.7.1 (subsection "strain hardening"), it was established that the ultimate moment capacity of columns is 2209 kNm. Therefore, the connections of columns to the foundation should be designed for 2209 kNm moment. It was also established that the ultimate moment capacity of beam at levels 1–3 is 1850 kNm and of beams at levels 4 and 5 is 1050 kNm. Therefore, the connections of beams to columns should be designed for 1850 kNm moment at levels 1–3 and for 1050 kNm moment at levels 4 and 5.

4.11   Uncertainty in the Strength of Structural Elements

There is some uncertainty in the moment capacity of beams and columns. Lower bound estimates of moment capacity should be used to calculate the deflections and plastic rotations. Upper bound estimates of moment capacity should be used to calculate the base shear, overturning moment, and connection forces.

4.12   Effect of Height on Building Vulnerability

Figure 4.28 compares the capacity curve of one-story building analyzed in Chap. 3
with the capacity curve of five-story building analyzed in this chapter. The pseudo-
acceleration is the normalized strength (lateral strength divided by the weight) of the
building. Due to its much higher weight, the five-story building has smaller normalized strength than the one-story building. However, the five-story building is much
more deformable than the one-story building because deformations in various stories
of the building are cumulative. The seismic toughness ST is the area below the
capacity curve. The seismic toughness of the five-story building is nearly same as
that for the one-story building. What a five-story building lacks in strength, it gains
in deformability. Therefore, the height of a building alone does not make it more
vulnerable to ground shaking. Low-rise buildings are better suited against low-
frequency ground motions on soft soil sites due to distant-large earthquakes. High-
rise buildings are better suited against high-frequency ground motions on rock sites
due to closer-small earthquakes.

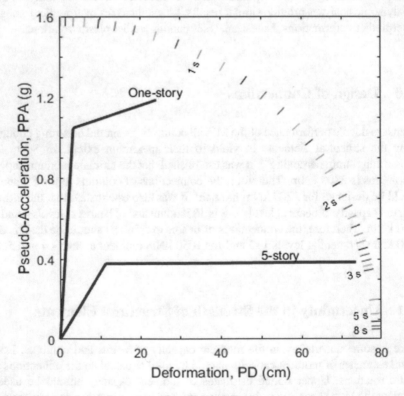

Fig. 4.28 Capacity curves for one- and five-story buildings

4.13   Effect of Soft-Story on Building Vulnerability

Figure 4.29 shows the sketch of a "soft-story" building. The MFs are infilled with masonry except at the lowest level which is used for parking or commercial purposes. Infills do not allow the upper story moment frames to deform freely. As a result, most of the deformation takes place in the first story. Figure 4.30 compares the capacity curve of a soft-story building with that of a regular MF building. Both buildings have similar normalized strength, but the soft-story building has much smaller deformability. The seismic toughness ST of the soft-story building is about half of that of the regular building. Therefore, a soft-story building is always more vulnerable to seismic shaking than a regular building.

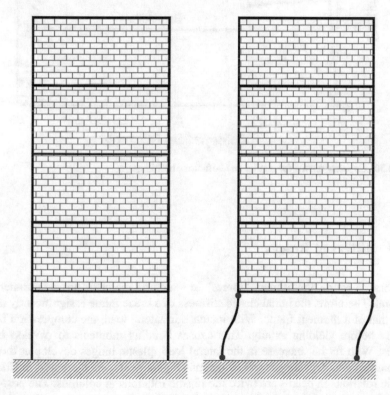

Fig. 4.29 Deformation of a soft-story building

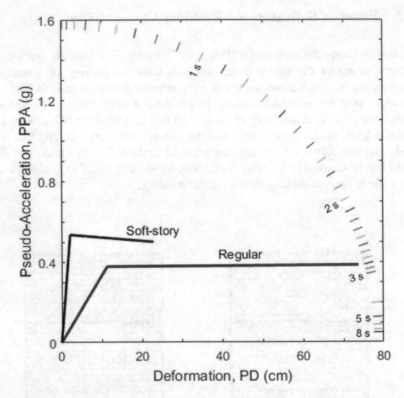

Fig. 4.30 Capacity curves for regular and soft-story buildings

4.14   Effect of Braces

In a braced frame, the primary "forces" in structural members are axial rather than flexural. Therefore, the initial elastic stiffness of a brace frame is significantly higher than that of a moment frame. With increase in lateral load, the compression braces buckle before yielding axially. This causes bending moments to develop in the beams. With further increase in the lateral load, plastic hinges develop at the base of columns and near the centers of beams (Figure 4.31). As discussed in Chap. 3, plastic rotations in beams are twice the plastic rotations in columns. The post-yield strength of a braced-frame building is similar to that of a moment-frame building, but the deformability of a braced-frame building is less due to large plastic rotations in the beams.

Fig. 4.31 Plastic hinges in a braced frame after buckling of compression braces

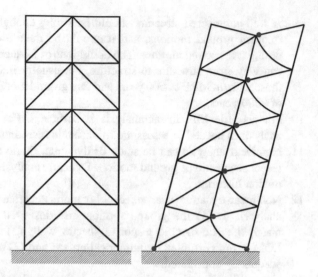

4.15 Summary

1. In general, a multistory MF has smaller normalized strength but more deformability than a one-story MF.
2. A multistory MF is ideal to resist high-frequency ground motions, such as those on rock sites due to closer-small earthquakes.
3. A single-story MF is ideal to resist low-frequency ground motions, such as those on soft-soil sites due to distant-large earthquakes.
4. Plastic yielding in beams can be a significant source of deformability and damping for a multistory MF.
5. Plastic yielding in columns can result in a "soft story," whereby deformations are concentrated in a single story. Soft story can be avoided by using strong columns and weak beams and by avoiding infills that restrict deformations in certain stories.
6. Plastic yielding does some damage to a moment frame, but limited plastic yielding can be tolerated as long as the structure meets its desired performance objective.
7. Plastic yielding in beams will occur only if the connections are stronger than the beams. Weak connections will fail before beams can yield, thus resulting in a brittle system. It is possible to have a brittle system made of ductile elements. The design strength of connections should account for strain hardening of structural members.
8. There is some uncertainty in the strength of structural members. Lower estimates of strength should be used to calculate the deflections and plastic rotations. Upper estimates of strength should be used to calculate the base shear, overturning moment, and connection forces.

9. A well-engineered structure should be strong enough to avoid damage during frequent ground motions, and it should be ductile enough to avoid collapse during rare ground motions. The definitions of frequent and rare ground motions can vary from structure to structure, but usually frequent ground motions are those with an MRP of 500 years and rare ground motions are those with an MRP of 2500 years.

10. For a brittle MF, linear analysis is sufficient. For a ductile MF, nonlinear analysis is needed to assess the full seismic resistance of the structure.

11. Nonlinear analysis can be static or dynamic. Static analysis makes use of the response spectra of ground motion. Dynamic analysis makes use of the ground motion histories.

12. Nonlinear-dynamic analysis can be more accurate than the nonlinear-static analysis, if: (1) the ground motion histories used in the dynamic analysis match the site-specific response spectra well; (2) the solution of nonlinear EOM is not corrupted by numerical errors; and (3) all sources of damping are adequately considered.

13. The results of nonlinear-dynamic analysis should not be blindly trusted because the analysis is prone to numerical errors. The results of nonlinear-dynamic analysis can also be corrupted by numerical damping used to obtain stable solutions. Implied damping in nonlinear-dynamic analysis should be checked. A nonlinear-dynamic analysis should always be preceded with a nonlinear-static analysis.

References

1. Hamburger, R. O., Krawinkler, H., Malley, J. O., & Adan, S. A. (2009, June). Seismic design of steel special moment frames: A guide for practicing engineers. *NIST GCR 09-917-3*, U.S. Department of Commerce, National Institute of Standards and Technology.
2. Engelhardt, M. D., Winneberger, T., Zekany, A. J., & Potyraj, T. J. (1996, August). *The Dogbone connection: Part II*. Modern Steel Construction.
3. Paulay, T. (1978). A consideration of p-delta effects in ductile reinforced concrete frames. *Bulletin of the New Zealand National Society for Earthquake Engineering., 111*(3), 151–160.
4. AISC. (2016). Seismic provisions for structural steel buildings. *ANSI/AISC 341-16*, American Institute of Steel Construction, Inc., Chicago, IL.
5. AISC. (2016). Prequalified connections for special and intermediate steel moment frames for seismic applications including supplement no. 1. *ANSI/AISC 358-16*, American Institute of Steel Construction, Inc., Chicago, IL.
6. ASCE. (2013). Seismic evaluation and retrofit of existing buildings. *ASCE 41-13*, American Society of Civil Engineers, Reston, VA.
7. Clough, R. W., & Penzien, J. (1976). *Dynamics of structures*. McGraw Hill Publications.
8. Chopra, A. K. (2017). *Dynamics of structures* (5th ed.). Pearson Publication.
9. CSI. (2011). *SAP 2000*. Berkeley, CA: Computers and Structures.
10. Malhotra, P. K. (2002). Cyclic-demand spectrum. *Earthquake Engineering and Structural Dynamics, 31*(7), 1441–1457.

Chapter 5
Sliding of Objects During Earthquakes

Nomenclature

ζ	Equivalent viscous damping
μ	Coefficient of dynamic (kinematic) friction
μ_s	Coefficient of static friction
$a_1(t), a_2(t)$	Accelerations of ground at time t in two orthogonal horizontal directions
$a_v(t)$	Vertical acceleration of ground at time t
cm	Centimeter
EOM	Equation(s) of motion
g	Acceleration due to gravity $= 9.81$ m/s^2
m	Mass of object
PGA	Peak horizontal ground acceleration
PGA_V	Peak vertical ground acceleration
PPA	Peak pseudo-acceleration (also known as spectral acceleration)
PD	Peak deformation (sometimes called spectral displacement)
PPV	Peak pseudo-velocity (also known as spectral velocity)
s	Second
t	Time
T	Undamped natural period
u	Sliding displacement

5.1 Introduction

Traditional seismic design relies on anchorage to resist seismic sliding. However, anchorage may not always be the best solution. For example, consider a broad piece of equipment resting on a concrete slab at a facility. The equipment is not anchored

© Springer Nature Switzerland AG 2021

P. K. Malhotra, *Seismic Analysis of Structures and Equipment*,
https://doi.org/10.1007/978-3-030-57858-9_5

to its foundation, either because: (1) there was no awareness of seismic risk when the facility was originally built or (2) the equipment is frequently moved within the facility to meet operational demands. The owner of the facility wants to reduce the seismic risk. There are two options: (1) anchor the equipment to its foundation or (2) let the equipment slide during earthquakes and provide flexibility in attached pipes, conduits, and ducts to accommodate sliding displacement.

If the first option is selected, the equipment will move with the ground during earthquakes; it will experience the same or higher acceleration than ground. The foundation will experience the inertial force (mass × acceleration) transmitted by the equipment. If the second option is selected, the equipment will slide when the inertial force approaches the frictional resistance between the equipment and the foundation. Therefore, the inertial force experienced by the equipment and the foundation is limited by the frictional resistance. However, any attachments to the equipment will have to absorb the sliding displacements. Whether the first or the second option is economical will depend on the following factors:

- Vulnerability of the equipment to accelerations
- Ability of attachments (pipes, conduits, and ducts) to absorb sliding displacements and/or the cost of retrofitting attachments to absorb sliding displacements
- Cost of anchoring the equipment and any business interruptions associated with installing anchors and
- Ability of the foundation to resist the seismic forces or the cost of strengthening the foundation to resist inertial forces transmitted by the equipment

To evaluate the above factors, the sliding response of the equipment needs to be computed efficiently and reliably. Seismic sliding has been the subject of several past studies [1–5]. Numerical solution of seismic sliding is well developed and understood [6, 7]. Sliding response of unanchored objects is of significant interest in the nuclear industry. ASCE 43-05 [8] standard for nuclear facilities includes a method for estimating the sliding response without performing a dynamic analysis, but the method is somewhat arbitrary and overly conservative.

The objectives of this chapter are to:

- Discuss nonlinear-dynamic analysis of sliding response
- Present nonlinear-static analysis of sliding response, and compare it with: (a) the nonlinear-dynamic analysis, and (b) the ASCE 43-05 [8] analysis and
- Present seismic sliding as an alternative design option

5.2   Sliding or Rocking

Some unanchored objects are prone to sliding while others are prone to rocking at their base. Figure 5.1 shows an unanchored object of mass m resting on its foundation. The center of gravity (CG) of the object is at height h from the base and distance b from the edge of the object. In Fig. 5.2, the foundation is shaken by time-varying,

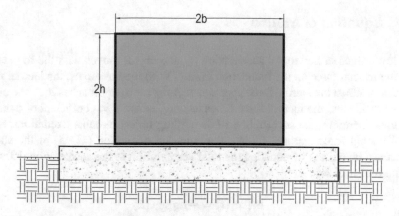

Fig. 5.1 An unanchored broad object resting on its foundation

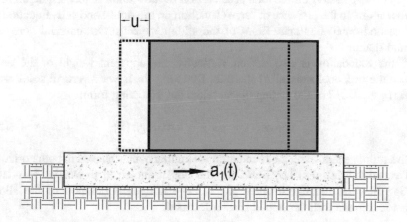

Fig. 5.2 Sliding of unanchored broad object due to horizontal shaking of its foundation

uniaxial, horizontal ground acceleration $a_1(t)$. The coefficient of static friction between the object and the foundation is μ_s. The maximum horizontal force that can be transmitted to the object is $\mu_s mg$, where g = acceleration due to gravity = 9.81 m/s^2. This force can be assumed to act at the CG of the object at height h. Hence, the moment trying to overturn the object is $\mu_s mgh$. The restoring moment due to the weight of the object is mgb. If mgb is greater than $\mu_s mgh$ (or $b/h > \mu_s$), the object will slide without rocking. If $b/h < \mu_s$, the object will rock without sliding. Slender objects are prone to rocking; broad objects are prone to sliding. Objects on smooth surfaces are prone to sliding; objects on rough surfaces are prone to rocking. This chapter is devoted only to sliding objects, for which $b/h > \mu_s$. Rocking objects are discussed in Chap. 6.

5.3 Equation of Motion

For low values of horizontal acceleration a_1, the object moves with the foundation and the inertial force on the foundation is ma_1. With increase in a_1, the inertial force increases. When the inertial force approaches $\mu_s mg$ (or a_1 approaches $\mu_s g$), the object starts to slide and the inertial force drops to μmg, where μ = coefficient of dynamic friction. Therefore, the acceleration of the sliding object remains fixed at μg. Since the frictional force opposes sliding, the equation of motion (EOM) of the sliding object can be written by equating the total acceleration to the acceleration delivered by the frictional force:

$$\ddot{u} + a_1(t) = -\mu g \, \text{sgn}\,(\dot{u}) \tag{5.1}$$

where u = sliding displacement (relative to the foundation); the over dots denote differentiation with respect to time t; and sgn = the "sign" function. Equation 5.1 is simply an expression for the total acceleration of the sliding object. Equation 5.1 is nonlinear due to the presence of "sign" function on the right-hand side. Equation 5.1 is a second-order, nonlinear EOM of the sliding object under uniaxial horizontal ground shaking.

If the foundation is also shaken vertically, the apparent weight of the object (hence the frictional resistance) changes. Denoting the upward vertical acceleration by $a_v(t)$, the EOM of the sliding object takes the following form:

$$\ddot{u} + a_1(t) = -\mu[g + a_v(t)] \, \text{sgn}\,(\dot{u}) \tag{5.2}$$

In a more general case, the foundation is simultaneously shaken in both horizontal and vertical directions. The EOM for the object subjected to triaxial foundation motion may be found elsewhere [6, 7]. The numerical solution of the sliding response is quite well developed and understood.

5.4 Ground Motion

In this chapter, the sliding response of an unanchored object is calculated during the 500-year MRP ground motion at a site in the San Francisco Bay Area (SFBA). The 500-year MRP ground motion for the SFBA site was determined in Chap. 2. Figure 5.3 shows the 500-year MRP response spectra for various values of damping. The response spectra are shown in the acceleration–deformation format. These are the demand curves for the site. The peak pseudo-acceleration PPA is read along the vertical axis and peak deformation PD is read along the horizontal axis. The natural period T is shown by the parallel diagonal lines. The PPA, PD, and T are related to each other by the expression: $PD = PPA \times (T/2\pi)^2$. Figure 5.4 shows one of the

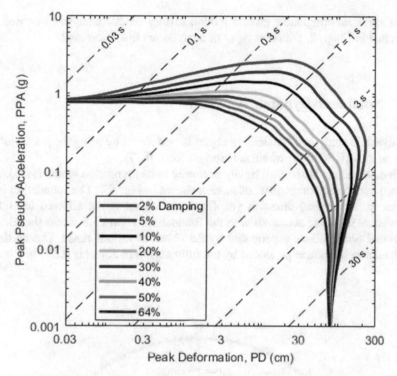

Fig. 5.3 500-year MRP ADRS for various values of damping for the SFBA site

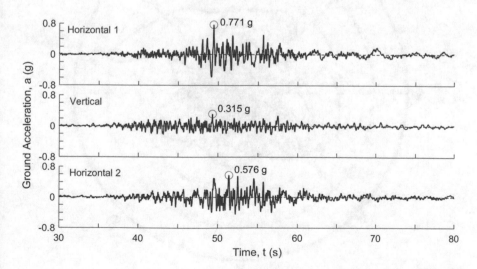

Fig. 5.4 One of seven 500-year MRP spectrum-compatible ground motion history for 3D analyses

seven spectrum-compatible ground motion history for 3D analyses; this was also generated in Chap. 2. Different types of analysis are discussed next.

5.5 Nonlinear-Dynamic Analysis

The dynamic analysis of the sliding object is performed by using the *frictional-link element* in SAP 2000 [9] which is based on Refs. [6, 7].

First, the object is assumed rigidly anchored to its foundation which is shaken by the two horizontal components of motion shown in Fig. 5.4. The foundation is not shaken in the vertical direction yet. Since the object is not allowed to slide, it experiences the same acceleration as the foundation. Figure 5.5 shows the resultant horizontal accelerations experienced by the object at various times. The maximum resultant acceleration experienced by the fully anchored object is 0.803 g.

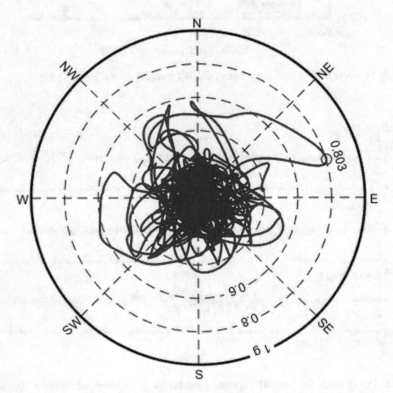

Fig. 5.5 Resultant horizontal accelerations experienced by a rigidly anchored object subjected to the two horizontal components of motion shown in Fig. 5.4

5.5.1 Response to Horizontal Shaking

Next, the object is assumed to simply rest on the foundation without any anchorage. Both the static and the dynamic coefficients of friction between the object and the foundation are assumed to be $\mu = 0.462$. This is 80% of the expected static coefficient of the friction between steel and concrete [10]. Since the maximum frictional resistance is μmg, the object slides as soon as the resultant horizontal ground acceleration reaches $\mu g = 0.462$ g. Figure 5.6 shows the resultant horizontal accelerations experienced by the object at various times. As expected, the acceleration never exceeds 0.462 g in any horizontal direction.

Figure 5.7 shows the sliding displacements of the object at various times. The maximum displacement is 4.97 cm and the residual displacement (at the end of ground shaking) is 4.96 cm. The sliding response of the object is computed for all seven simulations of the 500-year MRP ground motion. The second, third, and fourth columns in Table 5.1 list the peak values of acceleration, displacement, and residual displacement experienced by the object subjected to seven ground motions. The median values in the last row of Table 5.1 are the responses to the 500-year

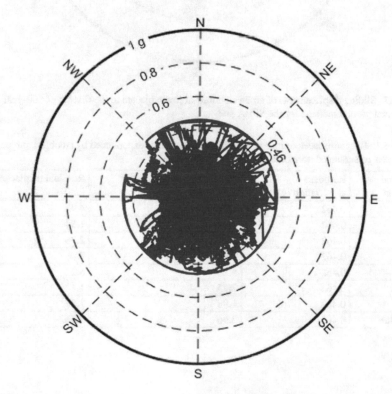

Fig. 5.6 Resultant horizontal accelerations experienced by an unanchored object subjected to the first set of 500-year MRP horizontal ground motions for the SFBA site

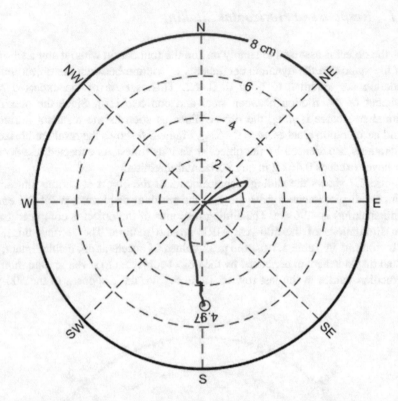

Fig. 5.7 Sliding displacements of an unanchored object subjected to the first set of 500-year MRP horizontal ground motions for the SFBA site

Table 5.1 Response accelerations and sliding displacements experienced by an object subjected to seven sets of simulated horizontal ground motions

Ground motion	Response acceleration (g)	Maximum sliding displacement (cm)	Residual displacement (cm)
1	0.462	4.97	4.96
2	0.462	1.91	1.54
3	0.462	2.96	2.76
4	0.462	7.66	7
5	0.462	5.26	5.22
6	0.462	2.75	1.1
7	0.462	1.88	1.87
Median	**0.462**	**2.96**	**2.76**

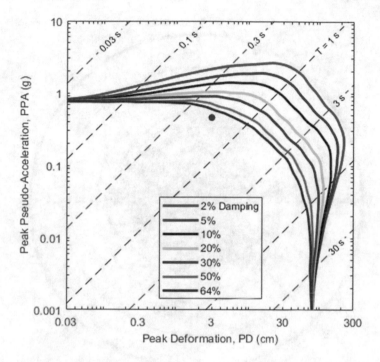

Fig. 5.8 Equilibrium point from dynamic analysis on the 500-year MRP response spectra for various values of damping

MRP ground motion at the SFBA site. The peak values of acceleration and displacement from dynamic analysis are 0.462 g and 2.96 cm, respectively. These are shown by a red dot on the response spectra plots in Fig. 5.8. Notice that the red dot is within the 64% damping response spectrum. Therefore, the implicit damping in dynamic analysis is greater than 64% of critical. From extrapolation, the damping is estimated at 92% of critical from dynamic analysis. So far, the effect of vertical ground motion has not been considered.

5.5.2 Effects of Vertical Ground Motion

The vertical ground motion affects the frictional resistance between the object and the foundation. The upward acceleration of the foundation increases the frictional resistance by increasing the "apparent weight" of the object. The downward acceleration decreases the frictional resistance by decreasing the "apparent weight" of the object.

Figure 5.9 shows horizontal accelerations experienced by the object when it is subjected to both horizontal and vertical ground motions. Note that the maximum acceleration experienced by the object is greater than 0.462 g. To understand this

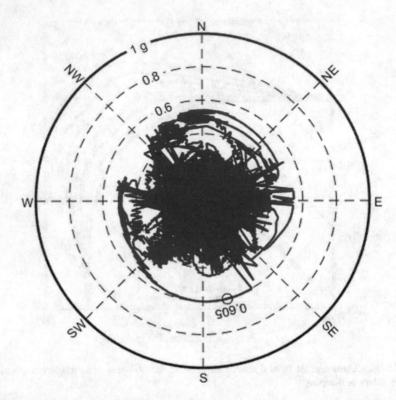

Fig. 5.9 Resultant horizontal accelerations experienced by an unanchored object subjected to the first set of 500-year MRP horizontal and vertical ground motions for the SFBA site

increase, recall that Eq. 5.2 is an expression for the total horizontal acceleration experienced by the object during sliding. In the absence of vertical motion, the maximum value of the right-hand side of Eq. 5.2 is μg. In the presence of vertical motion, the maximum value of the right-hand side is $\mu|g + a_v(t)|_{max}$, which is greater than μg. A conservative estimate of the maximum horizontal acceleration experienced by the object is $\mu(g + PGA_V)$, where PGA_V = peak ground acceleration in the vertical direction.

Figure 5.10 shows the sliding displacements of the object subjected to both horizontal and vertical ground motions. Compare this with Fig. 5.7, which shows sliding displacements due to horizontal shaking only. The sliding displacement of the object has increased in this case. The effect of vertical motion on sliding displacement was computed for all seven ground motions. Table 5.2 lists the responses from seven dynamic analyses. Note that in most cases, vertical ground motion increased the sliding displacement. This is probably because vertical ground motion makes it more difficult for the object to remain "attached" to the foundation.

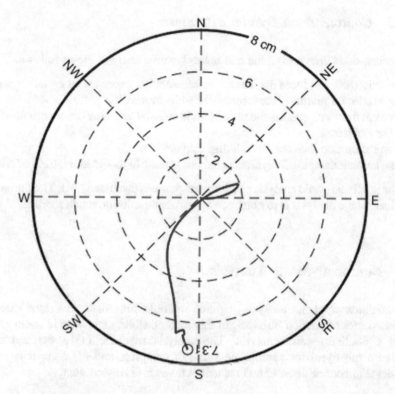

Fig. 5.10 Sliding displacements of an unanchored object subjected to the first set of 500-year MRP horizontal and vertical ground motions for the SFBA site

Table 5.2 Response accelerations and sliding displacements experienced by an object subjected to seven sets of simulated horizontal and vertical ground motions

Ground motion	Response acceleration (g)	Maximum sliding displacement (cm)	Residual displacement (cm)
1	0.605	7.31	7.3
2	0.573	2.72	2.01
3	0.586	2.5	0.392
4	0.544	8.84	8.45
5	0.533	7.83	7.78
6	0.538	3.27	1.01
7	0.591	4.4	4.38
Median	**0.573**	**4.4**	**4.38**

5.5.3   Comments on Dynamic Analysis

The conclusions drawn from the nonlinear-dynamic analysis are as follows:

- Seismic sliding reduces the maximum acceleration experienced by an object and the maximum inertial force transmitted to its foundation.
- Vertical motion increases the maximum horizontal acceleration experienced by a sliding object.
- Vertical motion increases the sliding displacement.
- The implicit damping in dynamic analysis is quite high—at about 92% of critical.

The nonlinear-static analysis of sliding response is discussed next. The purpose of nonlinear-static analysis is to obtain an approximate solution much more efficiently.

5.6   Nonlinear-Static Analysis

The nonlinear-dynamic analysis requires multiple runs using a set of carefully generated ground motion histories. In this section, sliding response is computed by using a nonlinear-static analysis. The analysis requires: (1) a demand curve (response spectrum) for appropriate value of damping, and (2) a capacity curve. The development of demand and capacity curves is discussed next.

5.6.1   Demand Curve

Figure 5.11a shows the cyclic force–displacement relationship for a sliding object. As the lateral force F on the object is gradually increased from zero, the object remains stationary until the force overcomes the maximum frictional resistance μmg and the object starts to slide. During sliding, the lateral force F remains fixed but the displacement increases. Upon reversing the direction of force, the object comes to rest and it does not move until the force reaches the frictional resistance in the opposite direction, i.e., $F = -\mu mg$. Figure 5.11a shows a complete force–displacement cycle consisting of loading, unloading, and reloading paths. The area enclosed within the force–displacement cycle in Figure 5.11a is the energy loss per cycle due to sliding. It is given by the following expression:

$$E_h = 4\mu mgD \tag{5.3}$$

The "strain energy" equals the area of the triangle in Fig. 5.11b:

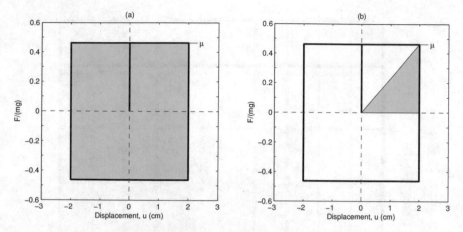

Fig. 5.11 Cyclic force–displacement relationship of a sliding object: (**a**) hysteretic energy lost by sliding; and (**b**) "potential energy" associated with sliding

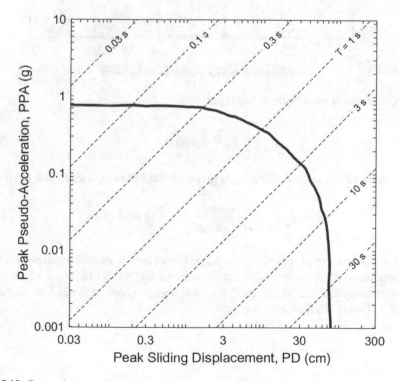

Fig. 5.12 Demand curve: 64% damping ADRS on logarithmic scale

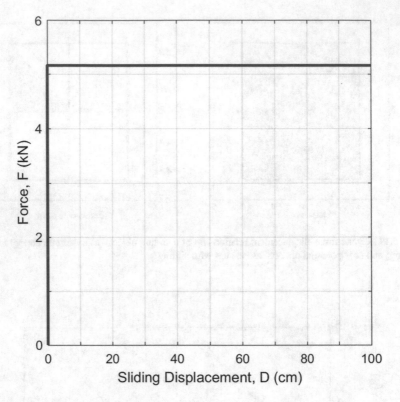

Fig. 5.13 Pushover curve for the sliding object

$$E_s = \frac{1}{2}\mu m g D \tag{5.4}$$

The equivalent-viscous damping is given by the following expression [11]:

$$\zeta = \frac{1}{4\pi}\frac{E_h}{E_S} = \frac{4\mu m g D}{4\pi\left(\frac{1}{2}\mu m g D\right)} = \frac{2}{\pi} = 0.64 \ (64\%) \tag{5.5}$$

Note that the damping for a rigid sliding object is 64% of critical, irrespective of the coefficient of friction μ and the amount of sliding D. Figure 5.12 shows a logarithmic plot of the ADRS for 64% damping; it is one of the curves shown in Fig. 5.3. This is the demand curve.

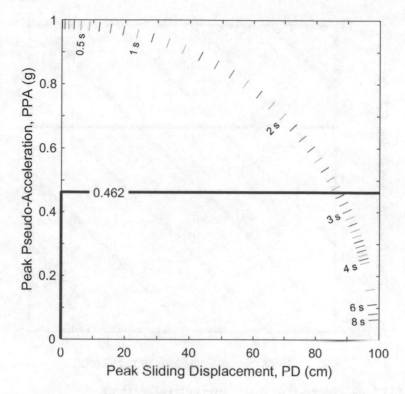

Fig. 5.14 Capacity curve for the sliding object

5.6.2 Capacity Curve

The capacity curve is generated from the pushover curve. The pushover curve is a plot between the lateral force and displacement; it is shown in Fig. 5.13. The object is assumed to have a mass of 1140 kg, the object remains at rest until the force reaches the frictional resistance, at which point the object starts sliding. The capacity curve is obtained by dividing the force along the vertical axis by the mass of the object. The capacity curve is shown in Fig. 5.14. A logarithmic plot of the capacity curve is shown in Fig. 5.15.

5.6.3 Sliding Response

The intersection between the demand curve and the capacity curve in Fig. 5.16 represents the equilibrium condition. According to the nonlinear-static analysis, the object will slide 5.96 cm during the 500-year MRP ground motion. The amount of sliding predicted by the nonlinear-static analysis is 35% higher than the value

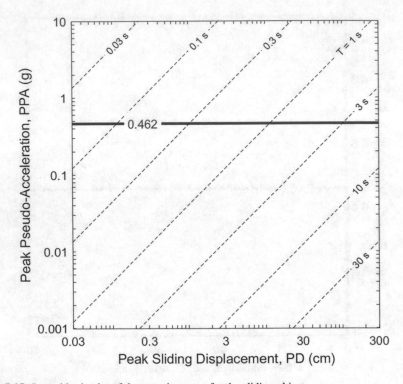

Fig. 5.15 Logarithmic plot of the capacity curve for the sliding object

obtained from the nonlinear-dynamic analysis without vertical motion. This much conservatism is considered appropriate for a much simpler approximate analysis which is not prone to numerical errors.

5.7 ASCE 43-05 Method

ASCE 43-05 [8] includes an approximate method, based on the "reserve energy" technique [12], for estimating the sliding displacement. The method consists of the following four steps:

1. Compute the effective coefficient of friction:

$$\mu_e = \mu(1 - 0.4PGA_V/g) \tag{5.6}$$

where, PGA_V = peak ground acceleration in the vertical direction. From Table 2.6, $PGA_V = 0.316$ g. Substituting, $\mu = 0.462$, and $PGA_V/g = 0.316$ into Eq. 5.6, the effective coefficient of friction is $\mu_e = 0.404$.

2. Compute the sliding coefficient:

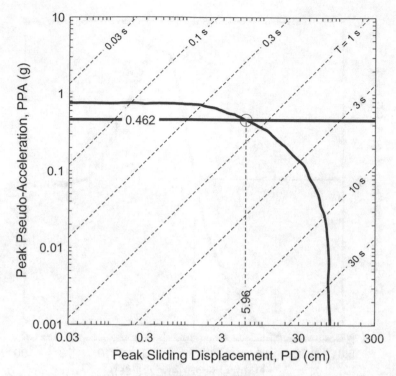

Fig. 5.16 Nonlinear-static analysis: intersection between demand curve (Fig. 5.12) and capacity curve (Fig. 5.15)

$$c_S = 2\mu_e g = 0.808 \text{ g} \tag{5.7}$$

3. Compute the lowest natural frequency f_{eS} at which the peak pseudo-acceleration *PPA* equals c_S on the 10% damping response spectrum. Figure 5.17 shows 10% damping response spectrum with natural frequency along the horizontal axis and peak pseudo-acceleration along the vertical axis. The lowest natural frequency at which $PPA = 0.808$ g is $f_{eS} = 0.745$ Hz.
4. Compute the sliding displacement as follows:

$$\delta_S = \frac{c_S}{(2\pi f_{eS})^2} = \frac{0.808 \times 981}{(2\pi \times 0.745)^2} = 36.2 \text{ cm} \tag{5.8}$$

The sliding displacement from ASCE 43-07 [8] method is eight times the value from the dynamic analysis and six times the value from the nonlinear-static analysis. Hence the ASCE 43-07 [8] method is overly conservative. The conservatism stems from: (1) arbitrary reduction in friction coefficient due to vertical motion (Eq. 5.6) and (2) arbitrary reduction in damping to 10% of critical.

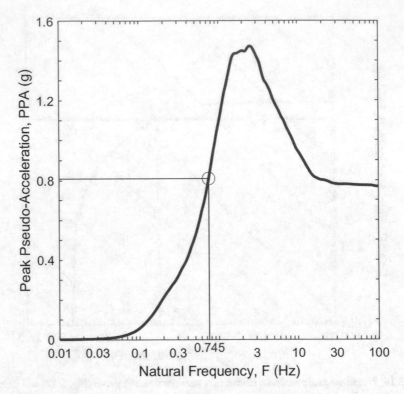

Fig. 5.17 Pseudo-acceleration response spectrum for 10% damping

5.8   Static and Dynamic Friction

The static coefficient of friction μ_s is about 25% higher than the dynamic (kinematic) coefficient of friction. The object needs to overcome the static friction before it can slide. The frictional resistance is also affected by the vertical ground shaking. The acceleration experienced by the object can be conservatively estimated from the following equation:

$$A = \mu_s(g + PGA_V) \tag{5.9}$$

The sliding displacements should be estimated from the dynamic (kinematic) coefficient of friction.

5.9   Sliding Response of Flexible Object

So far, the object has been assumed to be rigid. Real objects (structures and equipment) have some flexibility. Flexibility can significantly increase the acceleration experienced by an anchored object. However, flexibility does not affect the acceleration experienced by an unanchored object, because friction limits the force that can be transmitted to the object, whether it is rigid or flexible. Flexibility affects damping because not all the deformation is due to sliding; some deformation occurs in the object itself. Therefore, effective damping is expected to be smaller than $2/\pi$. In this section, the sliding response of a flexible object is computed by the nonlinear-static analysis. The capacity curve and the damping curve are generated as follows.

5.9.1   Capacity Curve

The object can be modeled as a lumped-mass system on a sliding base, as shown in Fig. 5.18. The object has a mass $m = 1000$ kg and a fixed-base period of $T = 0.5$ s. Therefore, the stiffness of the object is $k = m(2\pi/T)^2 = 1000(2\pi/0.5)^2 = 158$ kN/m. The coefficient of friction between the object and the foundation slab is assumed to be $\mu = 0.462$.

The pushover curve for the object is generated by determining the force required to deflect the mass by different amounts; it is shown in Fig. 5.19. For small deflections, the object deforms without sliding and the force increases linearly with deflection. When the force approaches the frictional resistance μmg, the object starts sliding without further increase in force. Dividing the force by the mass of the object gives the capacity curve. A linear plot of the capacity curve is shown in Fig. 5.20. The radial tick marks indicate the effective period of the object. The initial

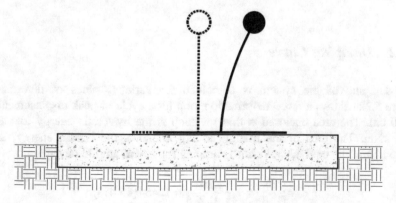

Fig. 5.18 Model of a flexible sliding object

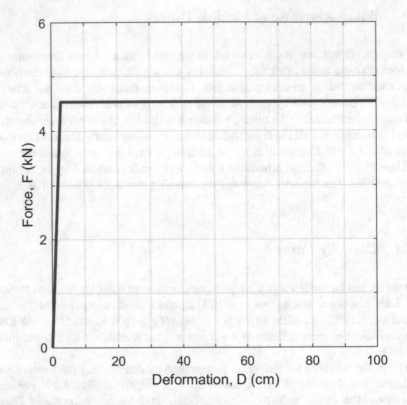

Fig. 5.19 Pushover curve for the flexible sliding object

period (before sliding) is $T = 0.5$ s. After sliding, the effective period continues to increase. Figure 5.21 shows a logarithmic plot of the capacity curve. The radial tick marks in Fig. 5.20 are replaced by parallel diagonal lines in Fig. 5.21.

5.9.2 Damping Curve

The damping of the system is computed for various values of deformation. Figure 5.22a shows a force–deformation loop for a cycle of peak displacement $PD = 10$ cm. The area enclosed within the loop is the hysteretic energy loss $E_h = 1.29$ kNm. The area of the triangle in Fig. 5.22b is the "strain energy" $E_s = 0.226$ kNm. The equivalent-viscous damping for 10-cm amplitude cycle is:

$$\zeta = \frac{E_h}{4\pi E_S} = \frac{1.29}{4\pi \cdot 0.226} = 0.454(45.4\%).$$

The damping is similarly computed for cycles of various amplitudes. Figure 5.23 shows a plot between peak deformation and damping. This is known as the damping

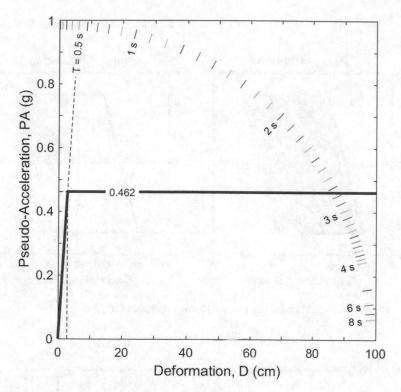

Fig. 5.20 Capacity curve for the flexible sliding object

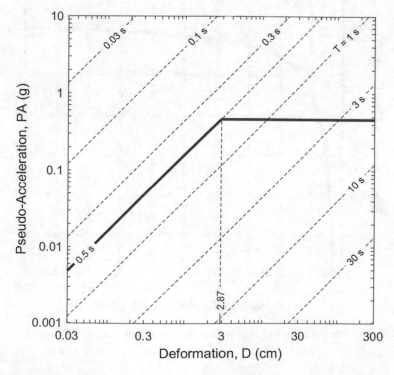

Fig. 5.21 Logarithmic plot of the capacity curve show in Fig. 5.20

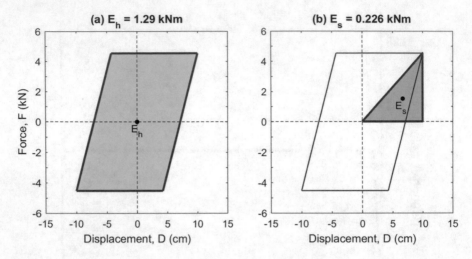

Fig. 5.22 Hysteretic and "strain" energies for 10-cm amplitude cycle

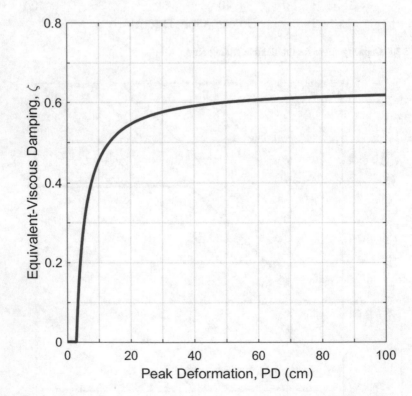

Fig. 5.23 Equivalent-viscous damping due to sliding for various values of deformation

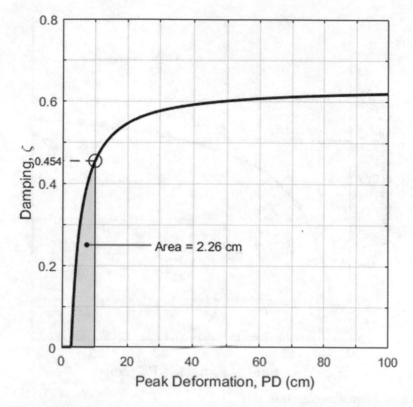

Fig. 5.24 Adjustment of damping curve

curve for the system. The damping curve of Fig. 5.23 needs some adjustments before it can be used in nonlinear-static analysis. Nonlinear-static analysis provides the peak response. During seismic shaking, the peak deformation occurs only once; rest of the times the deformation is less than the peak. Since the damping for smaller amplitude cycles is less, the damping is adjusted as follows. For $D = 10$ cm, the damping is 0.454 according to Fig. 5.23. The average (adjusted) damping for deformations between 0 and 10 cm is calculated by taking the area under the damping curve up to 10 cm and dividing that area by 10 cm (Fig. 5.24). This gives the average (adjusted) damping of $2.26/10 = 0.226$ (22.6%). The adjusted damping is similarly computed for other values of D. Finally, the damping is not allowed to drop below 2% of critical because there are some other sources of energy dissipation besides sliding. Figure 5.25 shows the adjusted damping curve for the object; this will be used to complete the nonlinear-static analysis.

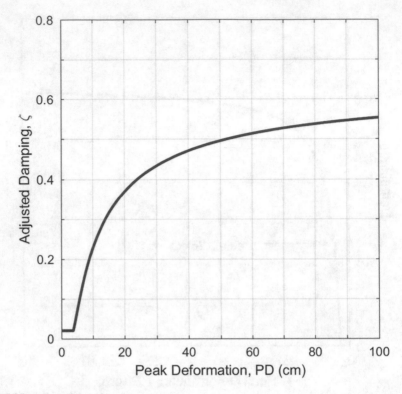

Fig. 5.25 Adjusted damping curve

5.9.3 Sliding Response

In Fig. 5.26, the capacity curve of Fig, 5.21 is superimposed on demand curves for various values of damping (Fig. 5.3). The intersections of capacity curve with demand curves provide deformations for various assumed values of damping. Figure 5.27 shows a plot of the deformation-versus-damping curve. Higher the damping, smaller is the deformation. But damping also depends on deformation as per the adjusted damping curve of Fig. 5.25. The intersection of deformation-versus-damping curve of Fig. 5.27 with the adjusted damping curve of Fig. 5.25 provides the damping and deformation at equilibrium (Fig. 5.28). The damping at equilibrium is 34% of critical and the deformation at equilibrium is 17.2 cm. Out of 17.2 cm deformation, 2.87 cm is elastic deformation of the object (Fig. 5.21) and the remaining 14.3 cm is due to sliding. The damping of flexible object is less than 64% because damping associated with elastic deformation of 2.87 cm is only 2% of critical.

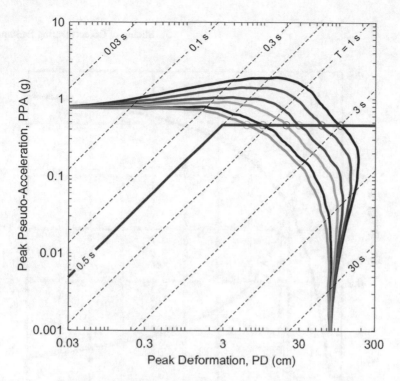

Fig. 5.26 Capacity curve superimposed on demand curves (response spectra) for various values of damping

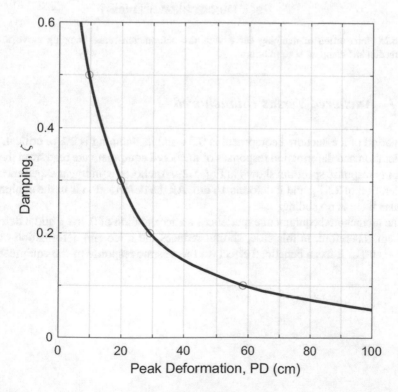

Fig. 5.27 Deformation-versus-damping curve for the flexible sliding object subjected to 500-year MRP ground shaking at the SFBA site

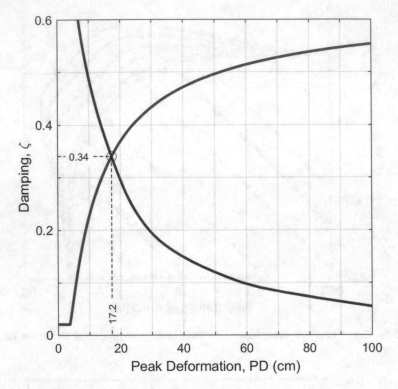

Fig. 5.28 Intersection of damping curve with the deformation-versus-damping curve to find deformation and damping at equilibrium

5.9.4 Anchored Versus Unanchored

The period of the anchored equipment is 0.5 s and its damping is 2% of critical. The acceleration and deformation responses of anchored equipment are read from the 2% damping response spectrum shown in Fig. 5.29. Anchored equipment experiences an acceleration of 2.57 g and it deforms 16 cm. All the deformation is in the equipment because there is no sliding.

The unanchored equipment experiences an acceleration of 0.462 g and it deforms 2.87 cm. Therefore, in this case, sliding reduces the force and deformation of the equipment, or it has a beneficial effect on the seismic response of the equipment.

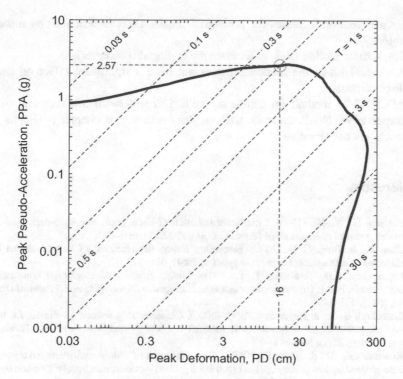

Fig. 5.29 Acceleration and deformation responses of anchored equipment

5.10 Effects of Foundation Tilt

The foundation may have some accidental tilt (slope). Tilt increases the tendency of the object to slide in the downward sloping direction. To study the effect of tilt, the foundation was assumed sloping by 2% (1.15°) in one direction and the sliding analysis was repeated. The sliding displacements were found to increase by less than 20%. Therefore, the effect of accidental tilt on sliding displacement was not considered significant because tilt greater than 1% is usually not tolerated.

5.11 Summary

1. Seismic sliding at the base of an object is a significant source of deformability and damping. Therefore, preventing seismic sliding is not always the best choice.
2. Seismic sliding limits the forces transmitted to the object and its foundation.

3. Vertical motion increases the horizontal accelerations experienced by a sliding object.
4. The vertical motion tends to increase the sliding displacement.
5. Accidental tilt in the foundation does not have a significant effect on sliding displacements.
6. ASCE 43-05 method for estimating sliding is somewhat arbitrary and overly conservative. Nonlinear-static analysis discussed in this chapter is simpler and much less conservative.

References

1. Shenton, H. W., III. (1996). Criteria for initiation of slide, rock, and slide-rock rigid-body modes. *Journal of Engineering Mechanics, ASCE, 122*(7), 690–693.
2. Choi, B., & Tung, C. D. (2002). Estimating sliding displacement of an unanchored body subjected to earthquake. *Earthquake Spectra, 18*(4), 601–613.
3. Lopez Garcia, D., & Soong, T. T. (2003). Sliding fragility of block-type non-structural components. Part 1: Unrestrained components. *Earthquake Engineering and Structural Dynamics, 32*(1), 111–129.
4. Chaudhuri, S. R., & Hutchinson, T. C. (2005). Characterizing frictional behavior for use in predicting the seismic response of unattached equipment. *Soil Dynamics and Earthquake Engineering, 25*(7), 591–604.
5. Konstantinidis, D., & Nikfar, F. (2015). Seismic response of sliding equipment and contents in base-isolated buildings subjected to broad-band ground motions. *Earthquake Engineering and Structural Dynamics, 44*(6), 865–887.
6. Park, Y. J., Wen, Y. K., & Ang, A. H.-S. (1986). Random vibration of hysteretic systems under bi-directional ground motions. *Earthquake Engineering and Structural Dynamics, 14*(4), 543–557.
7. Nagarajaiah, S., Reinhorn, A. M., & Constantinou, M. C. (1991). 3D-basis: nonlinear dynamic analysis of three-dimensional base isolated structures: Part II. *Technical Report NCEER-91-0005*, National Center for Earthquake Engineering Research, State University of New York at Buffalo, Buffalo, NY.
8. ASCE. (2005). Seismic design criteria for structures, systems and components in nuclear facilities. *ASCE 43-05*, American Society of Civil Engineers, Reston, VA.
9. Computers and Structures. (2011). *CSI analysis reference manual for SAP 2000*. Berkeley, CA: Computers and Structures, Inc..
10. Rabbat, B. G., & Russell, H. G. (1985). Friction coefficient of steel on concrete or grout. *Journal of Structural Engineering, American Society of Civil Engineers (ASCE), 111*(3), 505–515.
11. Chopra, A. K. (2011). *Dynamics of structures* (4th ed.). Prentice-Hall International Series in Civil Engineering and Engineering Mechanics.
12. Blume, J. A. (1960). A reserve energy technique for earthquake design and rating of structures in the inelastic range. In *Proceedings of 2nd world conference on earthquake engineering* (pp. 1061–1083). Tokyo: Science Council of Japan.

Chapter 6
Rocking of Objects During Earthquakes

Nomenclature

α	Critical base rotation (Fig. 6.1)
θ	Base rotation (Fig. 6.1)
$\dot{\theta}$	Angular velocity
$\ddot{\theta}$	Angular acceleration
θ_{max}	Maximum base rotation
ζ	Equivalent viscous damping (Eq. 6.11)
a	Aspect ratio $= b/h$
$a_H(t)$	Horizontal acceleration of ground at time t
$a_V(t)$	Vertical acceleration of ground at time t
b	Half-width of rectangular block
CG	Center of gravity
cm	Centimeter
EOM	Equation(s) of motion
g	9.81 m/s^2 = acceleration due to gravity
h	Height of CG (Fig. 6.1)
I	Moment of inertia (Eq. 6.4)
m	Mass of block
MRP	Mean return period
PGA	Peak horizontal ground acceleration
PGD	Peak horizontal ground displacement
PGV	Peak horizontal ground velocity
PPA	Peak pseudo-acceleration (also known as spectral acceleration)
PD	Peak deformation
PPV	Peak pseudo-velocity (also known as spectral velocity)
R	Distance between the point of rotation and CG (Fig. 6.1)
s	Second
t	Time
T	Undamped natural period

© Springer Nature Switzerland AG 2021
P. K. Malhotra, *Seismic Analysis of Structures and Equipment*,
https://doi.org/10.1007/978-3-030-57858-9_6

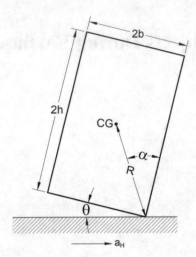

Fig. 6.1 A rigid rectangular object rocking at its base

6.1 Introduction

Rocking response of rigid rectangular objects was first studied by Housner [1]. He showed that smaller objects are more easily toppled than larger objects of the same aspect (width/height) ratio. Priestley et al. [2] emphasized the benefits of base rocking in reducing the forces transmitted to a structure during ground shaking. They used Housner's [1] results to estimate the rocking response by a nonlinear-static analysis. Yim et al. [3] performed a sensitivity study of rocking response using simulated ground motions. Makris and Konstantinidis [4] argued that the base rotations obtained from the analysis proposed by Priestly et al. [2] can be quite conservative. They presented a simplified form of the damping equation.

Rocking response of unanchored objects is particularly important in the nuclear industry. ASCE 43–05 [5] standard for nuclear facilities includes a method for estimating the rocking response without performing a dynamic analysis. This method is somewhat complex and has been shown to be unreliable by Dar et al. [6]. In the absence of a reliable static analysis, the only option is a dynamic analysis [4, 6]. Nonlinear-dynamic analyses are time-consuming because they require multiple runs using a suite of carefully selected ground motion histories. Also, nonlinear-dynamic analyses can be easily corrupted by numerical errors.

The objectives of this chapter are to: (1) discuss the nonlinear-dynamic and nonlinear-static analyses of rocking response and (2) present the concept of toppling response spectrum.

6.2   Equation of Motion

Figure 6.1 shows a rigid rectangular object of width $2b$ and height $2h$ shaken by horizontal ground acceleration $a_H(t)$ that is high enough to rock the object. The center of gravity (CG) of the object is assumed at its geometric center. The aspect ratio of the object is:

$$a = \frac{b}{h} \qquad (6.1)$$

When the base rotation θ reaches α (Fig. 6.1), the object is on the verge of toppling. Therefore, α is the critical base rotation; it is given by the following expression:

$$\alpha = \tan^{-1} a \qquad (6.2)$$

The radius of rotation (distance of CG from the point of rotation) is:

$$R = \sqrt{b^2 + h^2} = b\sqrt{1 + (1/a)^2} \qquad (6.3)$$

The mass of the object is m and the object's mass moment of inertia about the point of rotation is:

$$I = \frac{4}{3} mR^2 \qquad (6.4)$$

From the equilibrium of moments about the point of rotation, the equation of motion (EOM) of the object can be written as [4–6]:

$$I\ddot{\theta} + mgR \sin(\alpha \ \text{sgn}(\theta) - \theta) = -ma_H(t)R\cos(\alpha \ \text{sgn}(\theta) - \theta) \qquad (6.5)$$

In Eq. 6.5, the double dot above rotation θ denotes double differentiation with respect to time t; $g =$ acceleration due to gravity $= 9.81$ m/s^2; and "sgn" represents the sign function. The first term on the left side of Eq. 6.5 is the inertial moment and the second term is the restoring moment due to the weight of the object. The maximum absolute value of the restoring moment is mgb at $\theta = 0$ and its minimum value is zero at $\theta = \alpha$. The object will topple at $\theta = \alpha$ unless the ground moves fast enough to quickly reduce θ below α. Therefore, it is possible for the object to rotate slightly more than α without toppling.

The term on the right-hand side of Eq. 6.5 is the forcing function due to the horizontal acceleration of the ground. It is simply the mass of the object multiplied by the ground acceleration multiplied by the instantaneous height of the CG. The forcing function has some dependence on base rotation θ because the height of CG depends on (Fig. 6.1).

In Eq. 6.5, there is no separate term for damping (or energy dissipation). The energy is dissipated by impacts between the object and the base. Whenever the object strikes the base, its point of rotation changes from one corner to the other; the angular velocity of the object is multiplied by the coefficient of restitution, given by the following expression [1, 2]:

$$r = 1 - \frac{3}{2} \sin^2(\alpha) \qquad (6.6)$$

Housner [1] derived Eq. 6.6 by conserving the angular momentum when the point of rotation changes during impacts. Since $r < 1$, the angular velocity is reduced (or the energy is dissipated) by impacts. For slender objects, α is low; therefore, the coefficient of restitution ≈ 1. For stocky objects, α is high; therefore $r < 1$.

Dividing each term in Eq. 6.5 by I and substituting I from Eq. 6.4, yields:

$$\ddot{\theta} + \frac{3g}{4R} \sin\left(\alpha \ \text{sgn}\left(\theta\right) - \theta\right) = -a_H(t) \frac{3}{4R} \cos\left(\alpha \ \text{sgn}\left(\theta\right) - \theta\right) \qquad (6.7)$$

Equation 6.7 is a nonlinear, second-order differential equation. The solution of Eq. 6.7 depends on the ground acceleration history $a_H(t)$ and object's geometric characteristics α and R; α defines the shape (aspect ratio) of the object (Eqs. 6.1 and 6.2); and R defines the size of the object (Eq. 6.3).

6.3 Free-Vibration Response

The free-vibration response of the object is computed by ignoring the forcing function on the right-hand side of Eq. 6.7. The object is given an initial peak rotation *PR* and its response is computed at various times. There are two reasons to perform a free-vibration analysis: (1) to gain insight into the period and damping of the system and (2) to gain confidence in the numerical solution of the EOM.

The EOM of the object undergoing free vibration is:

$$\ddot{\theta} + \frac{3g}{4R} \sin\left(\alpha \ \text{sgn}\left(\theta\right) - \theta\right) = 0 \qquad (6.8)$$

Apart from the initial value of peak base rotation *PR*, the solution of Eq. 6.8 depends only on α and R.

Equation 6.8 is solved numerically for $\alpha = 0.3$ radian, $R = 1$ m, and $PR = 0.2$ radian by using Matlab [7] routine *ODE45*. Although, the coefficient of restitution, r is given by Eq. 6.6, its value is assumed equal to 1. In other words, the angular velocity $\dot{\theta}$ is not reduced after base impacts. Figure 6.2 shows plot of θ versus time t for $\alpha = 0.3$ radian, $R = 1$ m, and peak rotation $PR = 0.2$ radian. Note that the amplitude of oscillation does not reduce from one cycle to the next because the

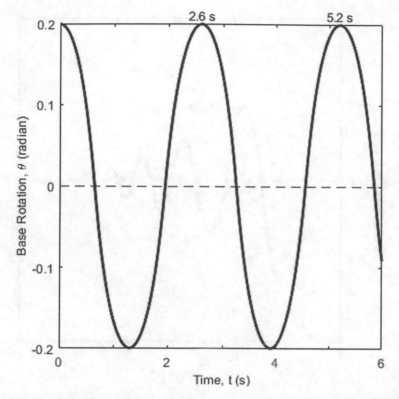

Fig. 6.2 Undamped free-vibration response for $\alpha = 0.3$ radian, $R = 1$ m, and $PR = 0.2$ radian

coefficient of restitution r has been assumed equal to 1. Since the system has no other way of losing energy, the amplitude of oscillations remains fixed at $PR = 0.2$ radian. The period obtained from the numerical analysis matches the period given by the following equation proposed by Housner [1]:

$$T = 8\sqrt{\frac{R}{3g}}\cosh^{-1}\left(\frac{1}{1 - PR/\alpha}\right) \qquad (6.9)$$

Therefore, Eq. 6.9 [1] provides a good estimate of the undamped natural period of the system. Also, the numerical solution of the equation of motion by the Matlab [7] routine *ODE45* can be trusted as long as the error tolerances during the solution are kept sufficiently low. According to Eq. 6.9, the period of the rocking object is zero when the amplitude of oscillation approaches zero ($PR = 0$) and it is infinite when the amplitude of oscillation approaches critical angle ($PR = \alpha$).

Next, the free-vibration analysis is performed by using the "actual" value of the coefficient of restitution r given by Eq. 6.6. During the analysis, the base rotation θ and the angular velocity $\dot{\theta}$ are closely monitored. Whenever the base rotation

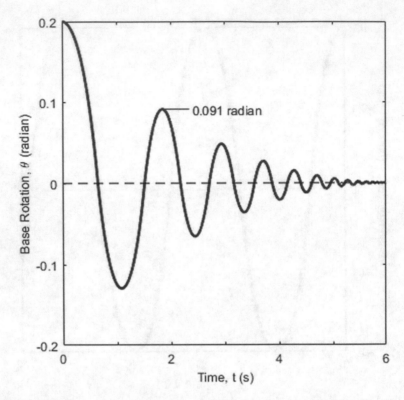

Fig. 6.3 Damped free-vibration response for $\alpha = 0.3$ radian, $R = 1$ m, and $PR = 0.2$ radian

changes sign (or an impact occurs), the angular velocity is multiplied by r to account for energy dissipation through impact. Figure 6.3 shows a plot of θ versus time t for $\alpha = 0.3$ radian, $R = 1$ m, and initial peak base rotation $PR = 0.2$ radian. As expected, the amplitude of oscillation reduces after every half-cycle because the system loses energy through base impacts. After the first cycle, the amplitude is 45.5% of the original value. Although the system is not viscously damped, the equivalent-viscous damping can be estimated from the rate of amplitude decay [8]:

$$\zeta = \frac{1}{2\pi} \ln \left(\frac{PR}{\theta_1} \right) \qquad (6.10)$$

where $\theta_1 =$ amplitude after one cycle. As reported in past studies [2, 4, 6], the equivalent-viscous damping is found to depend only on r, which depends on α (Eq. 6.6). The following expression by Makris and Konstantinidis [4] provides a good estimate of the equivalent-viscous damping:

$$\zeta = -0.68 \ln \left(1 - \frac{3}{2} \sin^2 \alpha\right) \tag{6.11}$$

The conclusions drawn from the free-vibration analyses are as follows:

- The undamped period of the system depends on R and the normalized amplitude of oscillation PR/α. Equation 6.9 by Housner [1] provides a good estimate of the period of the system.
- The equivalent-viscous damping of the system depends only on α. It does not depend on R or the amplitude of oscillation PR. Equation 6.11 by Makris and Konstantinidis [4] provides a good estimate of damping.

6.4   Ground Motion

In this chapter, the rocking response of the object is calculated during the 500-year MRP ground motion at a site in the San Francisco Bay Area (SFBA). The 500-year MRP ground motion for the SFBA site was determined in Chap. 2. Figure 6.4 shows the 500-year MRP response spectra for various values of damping. The response

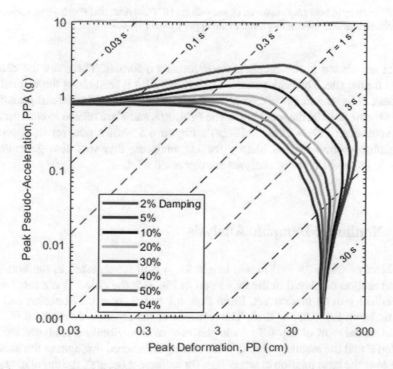

Fig. 6.4 500-year MRP ADRS for various values of damping for the SFBA site

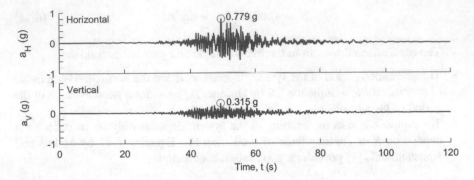

Fig. 6.5 One set of 500-year MRP ground motion history for the SFBA site

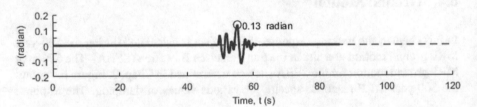

Fig. 6.6 History of base rotation for an object with $\alpha = 0.322$ radian, and $R = 3.16$ m, subjected to the horizontal ground motion displayed in the upper part of Fig. 6.5

spectra are shown in the acceleration–deformation format. These are the demand curves for the site. The peak pseudo-acceleration *PPA* is read along the vertical axis and peak deformation *PD* is read along the horizontal axis. The natural period *T* is shown by the parallel diagonal lines. The *PPA*, *PD*, and *T* are related to each other by the expression: $PD = PPA \times (T/2\pi)^2$. Figure 6.5 shows one set of spectrum-compatible ground motion history for 2D analyses; this was also generated in Chap. 2. Different types of analyses are discussed next.

6.5 Nonlinear-Dynamic Analysis

An object of width $2b = 2$ m and height $2h = 6$ m is subjected to the horizontal ground motion displayed in the upper part of Fig. 6.5; the object is not subjected to the vertical ground motion yet. From Eqs. 6.1 to 6.3, $\alpha = 0.322$ radian and $R = 3.16$ m. From Eq. 6.6, $r = 0.85$. The response of the object is determined from the numerical solution of Eq. 6.7. As in the case of free-vibration analysis, the base rotation θ and the angular velocity $\dot{\theta}$ are closely monitored throughout the analysis. Whenever the base rotation changes sign (or an impact occurs), the angular velocity is multiplied by r to account for energy dissipation through impact. Figure 6.6 shows

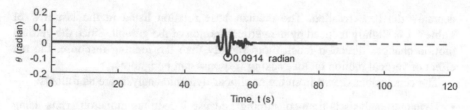

Fig. 6.7 History of base rotation for an object with $\alpha = 0.322$ radian, and $R = 3.16$ m, subjected to both horizontal and vertical ground motions displayed in Fig. 6.5

Table 6.1 Maximum base rotations for an object with $\alpha = 0.322$ radian, and $R = 3.16$ m, subjected to seven sets of simulated ground motions

	Peak base rotation PR (radian)	
Ground motion	Horizontal shaking only	Horizontal and vertical shaking
1	0.130	0.0914
2	0.196	0.146
3	0.197	0.216
4	0.0759	0.0788
5	0.198	0.153
6	0.151	0.157
7	0.276	0.242
Median	**0.196**	**0.153**

a plot of base rotation versus time. As can be seen in Fig. 6.6, the object does not topple; the base rotation returns to zero (or the object returns to its upright position) after the earthquake.

The vertical motion of the ground affects the restoring moment about the point of rotation. The restoring moment increases when the vertical acceleration is positive (upward), and the restoring moment decreases when the vertical acceleration is negative (downward). The EOM of the object under both horizontal and vertical ground motions is as follows [5]:

$$\ddot{\theta} + \frac{3g}{4R}\left(1 + \frac{a_V(t)}{g}\right)\sin\left(\alpha \ \mathrm{sgn}\left(\theta\right) - \theta\right) = -\frac{3}{4R}a_H(t)\cos\left(\alpha \ \mathrm{sgn}\left(\theta\right) - \theta\right) \quad (6.13)$$

where $a_V(t)=$ upward acceleration at time t. Equation 6.13 is solved using both horizontal and vertical components of the ground motion displayed in Fig. 6.5. Figure 6.7 shows a plot of base rotation versus time. Compare this with the plot in Fig. 6.6, which was generated without the vertical motion. The vertical motion has slightly reduced the maximum base rotation. The effect of vertical motion was considered for all seven simulated ground motions. In Table 6.1, the second column lists the base rotations due to horizontal motion only and the third column lists the base rotations due to both horizontal and vertical motions. Upon comparing the values in the second and third columns, it is seen that vertical motion can increase or decrease the base rotation. The median base rotation listed in the last row of Table 6.1 is slightly reduced by the vertical motion of the ground. Since the vertical motion changes direction much more rapidly than the rocking response, the net effect of vertical motion on the rocking response can be ignored.

The conclusions drawn from the nonlinear-dynamic analysis are as follows:

- Dynamic analysis is time-consuming because it requires numerous runs using several sets of simulated ground motions.
- Results of dynamic analysis can be easily corrupted by numerical errors. Therefore, extra care is needed to minimize numerical errors during dynamic analysis.
- The effect of vertical motion on the rocking response can be neglected.

The nonlinear-static analysis is discussed next. The purpose of nonlinear-static analysis is to obtain an efficient solution without numerical difficulties. The improvements made to overcome the shortcomings of previous nonlinear-static analyses [2] are identified.

6.6   Nonlinear-Static Analysis

Nonlinear-static analysis is an approximate solution of the EOM (Eq. 6.7). The right side of Eq. 6.7 represents the demand imposed by the ground motion and the left side represents the capacity of the system. The nonlinear-static analysis of the object of width $2b = 2$ m, and height $2h = 6$ m, subjected to the 500-year MRP ground motion at the SFBA site, is carried out as follows:

1. Compute a, α, and R from Eqs. 6.1, 6.2, and 6.3, respectively:

$$a = \frac{1}{3} = 0.333$$

$$\alpha = \tan^{-1} 0.333 = 0.322 \text{ radian}$$

$$R = \sqrt{1^2 + 3^2} = 3.16 \text{ m}$$

2. Substitute α into Eq. 6.11 to compute damping:
   $$\zeta = -0.68 \ln \left(1 - \tfrac{3}{2} \sin^2 0.322\right) = 0.111 \ (11.1\% \text{ of critical})$$
3. Generate 11.1% damping response spectrum of horizontal ground motion by interpolating between the 10% and 15% damping response spectra shown in Fig. 6.4. Figure 6.8 shows the 11.1% damping ADRS on a logarithmic scale. This is not the demand curve yet, because the ground acceleration on the right side of Eq. 6.7 is multiplied by a few factors.
4. Taking a cue from the right-hand side of Eq. 6.7, multiply PD by $3/(4R)$ to obtain peak rotation PR and multiply PPA by $3/(4R)$ to obtain peak angular acceleration PAA. Figure 6.9 shows the angular–acceleration–rotation plot of the response spectrum.

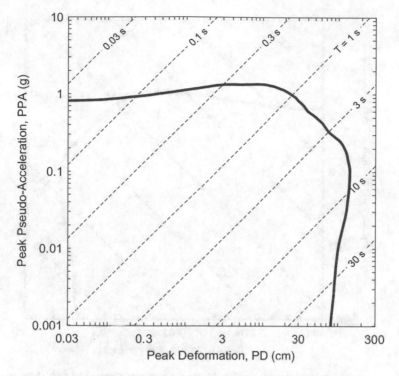

Fig. 6.8 11.1% damping ADRS for 500-year MRP ground motion at SFBA site

5. Once again, taking a cue from the right side of Eq. 6.7, multiply PR along the horizontal axis of Fig. 6.9 by $\cos(\alpha - PR)$ for all values of $PR \leq \alpha$. This correction accounts for the fact that the forcing function on the right side of Eq. 6.7 depends on the base rotation. The resulting plot is shown in Fig. 6.10. This is the demand curve for the site.

6. For different values of peak rotation PR between 0 and α, compute the natural period T from Eq. 6.9. Figure 6.11 shows a plot between the peak rotation PR and natural period T. For each pair of PR and T values, obtain peak angular–acceleration $PAA = PR(2\pi/T)^2$. Figure 6.12 shows a plot between PAA and PR. This is the capacity curve for the object.

7. Superimpose the capacity curve (Fig. 6.12) over the demand curve (Fig. 6.10). The two curves intersect at a rotation of 0.31 radian (Fig. 6.13). Therefore, the base rotation for the object during the 500-year MRP ground motion is 0.31 radian. Sometimes, the demand and capacity curves may intersect at more than one point. Since the rotation starts from zero, the intersection corresponding to the lowest value of rotation represents the true equilibrium condition, because the object will arrive at that rotation first and get "locked."

8. If the demand curve completely enveloped the capacity curve (without intersecting it), the object would be toppled by the ground motion. If the capacity

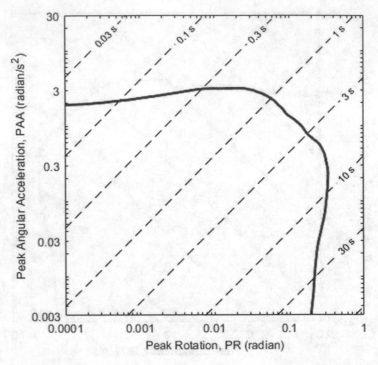

Fig. 6.9 Angular acceleration–rotation plot of response spectrum obtained by multiplying both *PD* and *PPA* in Fig. 6.8 by $3/(4R)$

curve completely enveloped the demand curve (without intersecting it), the object will not rock during the ground motion.

The results from the nonlinear-static analysis are somewhat conservative compared to those from the nonlinear-dynamic analysis, but they are free of numerical errors. Next, the improvements made in the proposed nonlinear-static analysis are highlighted.

The nonlinear-static analysis described in this chapter differs from the analysis proposed by Priestly et al. [2] in the following respects:

- In the Priestly et al. [2] analysis, the rotational spectrum is not computed. The deformation read from the deformation spectrum of horizontal ground motion is divided by h to compute rotation. In the method proposed in this chapter, the rotational spectrum is computed by multiplying the deformation spectrum by $3/(4R)$. Since $h < 4R/3$, the results from the Priestly et al. [2] analysis are conservative. Note that the $3/(4R)$ factor appears on the right-hand side of Eq. 6.7 because the mass is "rotational" while the excitation is "translational." If the mass were translational, the $3/(4R)$ factor would not be necessary.
- The correction to the rotational spectrum by $\cos(\alpha - PR)$ factor (Step 5) was omitted in the analysis by Priestly et al. [2]. However, the influence of this

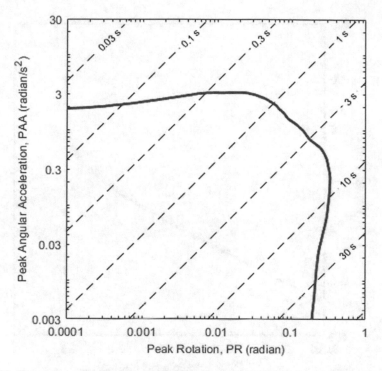

Fig. 6.10 Demand curve obtained by multiplying *PR* in Fig. 6.9 by $\cos(\alpha - PR)$ for $PR \le \alpha$

correction is not significant as can be seen from the comparison between Figs. 6.9 and 6.10.

- In the Priestly et al. [2] analysis, the equilibrium rotation was obtained iteratively by assuming a value of base rotation and then correcting it in successive itera- tions. As described under Step 7, multiple equilibrium conditions are possible but the true equilibrium condition corresponds to the lowest rotation. However, the Priestly et al. [2] solution could converge to any of the possible equilibrium conditions thus adding significant conservatism and uncertainty to the predicted response. The graphical solution presented in this chapter avoids this problem in addition to being efficient and transparent.

6.7   Safety Margin against Toppling

In the previous example, the rotation of the object under 500-year MRP ground shaking is 0.31 radian. Since critical rotation for the object is $\alpha = 0.322$ radian, it is tempting to say that the safety margin against toppling is $0.322/0.31 = 1.04$. How- ever, this can be misleading because rocking response is highly nonlinear and a specific increase in the ground motion does not produce a proportional increase in

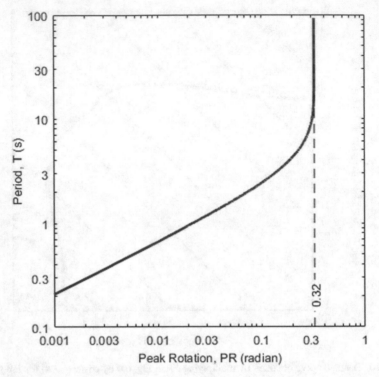

Fig. 6.11 Period T versus peak rotation PR for an object of width $2b = 2$ m and height $2h = 6$ m

the response [4]. Figure 6.14 shows that a 70% increase in the demand curve (response spectrum) is required for the demand curve to completely envelop the capacity curve, thus causing the object to topple. Therefore, the safety margin against collapse is 1.7 rather than 1.04. The 500-year MRP response spectrum multiplied by 1.7 is roughly the 2200-year MRP response spectrum for the SFBA site. It is most informative to say that the object is toppled by the 2200-year MRP ground motion for the site. The safety margin is best expressed in terms of the MRP of ground motion that will topple the object.

6.8   Toppling Response Spectrum

Between two objects of the same aspect ratio but different size, the smaller object is more easily toppled by the ground motion than the larger object. That is because, for toppling to occur, the CG of the larger object has to move more than that of the smaller object. Among objects of same width but different heights, the taller object is more easily toppled than the shorter object. The toppling heights were calculated for objects of various widths. A plot of toppling height versus width is shown in

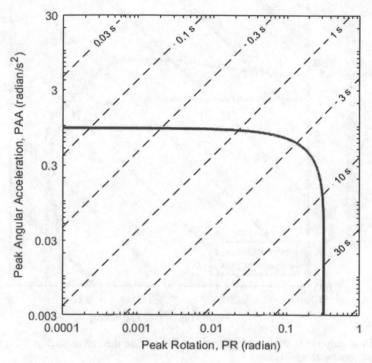

Fig. 6.12 Capacity curve obtained from Fig. 6.11 by using the relation $PAA = PR(2\pi/T)^2$

Fig. 6.15. This is the 500-year MRP toppling response spectrum for the SFBA site. No matter how tall the object, if it is more than 1.2 m wide, it will not be toppled by the 500-year MRP ground motion at SFBA site. This is because the 500-year MRP ground motion at the SFBA site lacks the displacement necessary to move the CG more than 0.6 m. Minimum acceleration is needed to rock the object, but minimum displacement is needed to topple the object.

Toppling response spectrum provides a practical way of assessing the toppling vulnerability of numerous objects at a site. For each object, based on its height and width, a point is marked on the toppling response spectrum. If the point is in the "toppled" zone, the object will be toppled by the 500-year MRP ground motion. The toppling risk can be mitigated by augmenting the width of the object, anchoring the object, or by providing lateral supports to the object. For objects at various floors of a building, toppling response spectra can be generated from the estimates of motions at various floors of a building. Toppling response spectra can of course be generated for different MRPs.

When an object is on the verge of toppling, there may be some other forces (e.g., wind, traffic, etc.) which may cause it to topple prematurely. Therefore, it is conservatively assumed that the object is toppled if during ground shaking the effective period of rocking response exceeds 10 s. With this definition, the object analyzed in Sect. 6.6 will be toppled because the period of rocking response at

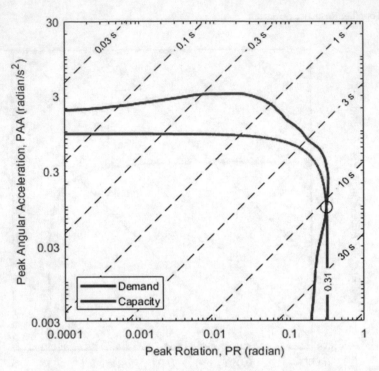

Fig. 6.13 Superposition of capacity curve (Fig. 6.12) over the demand curve (Fig. 6.10) to obtain peak base rotation at equilibrium

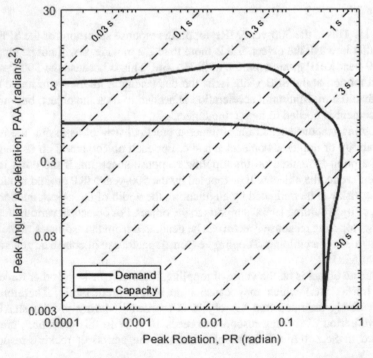

Fig. 6.14 Effect of 70% increase in the demand curve (response spectrum) on the rocking response of the object

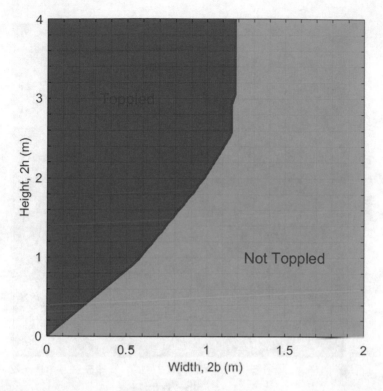

Fig. 6.15 Toppling response spectrum of the 500-year MRP ground motion at the SFBA site

equilibrium is slightly longer than 10 s, as seen in Fig. 6.13. Figure 6.16 shows a conservative plot of the toppling response spectrum.

6.9 Summary

1. The effective period of oscillation of a rocking object depends on: (a) its size and (b) the amplitude of rotation as a fraction of the critical rotation. The equation proposed by Housner [1] provides an accurate estimate of the effective period.
2. The equivalent-viscous damping of a rocking object depends on its aspect ratio. The equation proposed by Makris and Konstantinidis [4] provides a good estimate of the equivalent-viscous damping.
3. Vertical motion of the ground does not have a significant effect on the rocking response. Vertical motion changes direction much more rapidly compared to the rocking response. Therefore, the net effect of vertical motion on the rocking response can be neglected.
4. The response of a rocking object is highly nonlinear. An increase in the amplitude of ground shaking disproportionately increases the rocking response. The safety

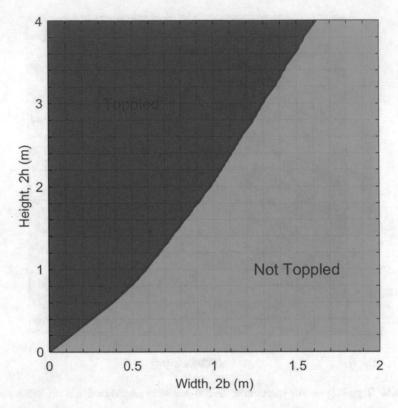

Fig. 6.16 Conservative plot of the toppling response spectrum of the 500-year MRP ground motion at the SFBA site

margin against toppling should not be expressed by the ratio between the critical rotation and the computed rotation. It is more meaningful to express the safety margin by the MRP of ground motion that will topple the object.

5. The nonlinear-static analysis presented in this study overcomes the shortcomings of the previous nonlinear-static analyses. Therefore, the results are not overly conservative compared to those from the nonlinear-dynamic analysis. In addition to being highly efficient, the analysis is not prone to numerical errors that can corrupt a nonlinear-dynamic analysis. Even when a nonlinear-dynamic analysis is performed, a nonlinear-static analysis should be performed first.

6. The toppling response spectrum has a simple form. It presents the minimum heights necessary to topple unanchored objects of various widths. Therefore, it is a practical way of assessing and mitigating toppling risk at a facility. No matter how high the PGA or how tall an object, it cannot be toppled unless its aspect ratio $b/h < 3/4(PGA/g)$, and its width $2b < 1.5PGD$.

7. A minimum PGA is needed to rock an object, but a minimum PGD is necessary to topple the object.

References

1. Housner, G. W. (1963). The behavior of inverted pendulum structures during earthquakes. *Bulletin of the Seismological Society of America, 53*(2), 403–417.
2. Priestley, M. J. N., Evison, R. J., & Carr, A. J. (1978). Seismic response of structures free to rock on their foundations. *Bulletin of the New Zealand National Society for Earthquake Engineering, 11*(3), 141–150.
3. Yim, C. K., Chopra, A., & Penzien, J. (1980). Rocking response of rigid blocks to earthquakes. *Earthquake Engineering and Structural Dynamics, 8*(6), 565–587.
4. Makris, N., & Konstantinidis, D. (2003). The rocking Spectrum and the limitations of practical design methodologies. *Earthquake Engineering and Structural Dynamics, 32*(2), 265–289.
5. ASCE. (2005). Seismic design criteria for structures, systems and components in nuclear facilities. In *ASCE 43–05*. Reston, VA: American Society of Civil Engineers.
6. Dar, A., Konstantinidis, D., & El-Dakhakhni, W. W. (2016). Evaluation of ASCE 43-05 seismic design criteria for rocking objects in nuclear facilities. *Journal of Structural Engineering, ASCE.*
7. MathWorks. (2020). *MATLAB Version 9.8.0.1417392 (R2020a)*. Natick, MA: The MathWorks, Inc.
8. Chopra, A. K. (2011). *Dynamics of Structures,* 4th edition, Prentice-Hall International Series in Civil Engineering and Engineering Mechanics.

Chapter 7
Seismic Response of Storage Racks

Nomenclature

θ	Joint rotation
ζ	Equivalent viscous damping ratio
ζ_h	Hysteretic damping
CA	Cross-aisle
CG	Center of gravity
DA	Down-aisle
E_h	Hysteretic energy lost per cycle
E_s	Strain energy
H	Height of CG
m	Total mass of rack and product
g	$9.81 \text{ m/s}^2 =$ acceleration due to gravity
M	Joint moment
MF	Moment frame(s)
MRP	Mean return period
OTM	Overturning base moment
PD	Peak deformation
PPA	Peak pseudo-acceleration
Q	Base shear
RMI	Rack Manufacturers Institute
SFBA	San Francisco Bay Area
T_{CA}	Natural period of rack in cross-aisle direction (before sliding)
W	Total weight of rack and product

© Springer Nature Switzerland AG 2021
P. K. Malhotra, *Seismic Analysis of Structures and Equipment*,
https://doi.org/10.1007/978-3-030-57858-9_7

7.1   Introduction

Storage racks are essential part of industrial and commercial facilities. They are often used to store high-value or hazardous products. During past earthquakes, products have fallen from shelves or the racks have collapsed. Failure of racks poses significant risk to life and property. Storage racks are generally made of cold-formed steel components. The columns (or uprights) are thin-walled open sections and beams are open or boxed sections. Columns have holes at regular interval to enable connections with beams at different heights (levels). The beams support the shelves.

In the cross-aisle (CA) direction, the lateral loads are resisted by trusses. In the down-aisle (DA) direction, trusses cannot be used because clear openings are needed to load and unload racks. Therefore, the lateral loads in the DA direction are resisted by moment frames. The beam-column moment connections are proprietary bolted (or pinned) connections which can be easily assembled or unassembled to achieve desired configuration of racks. The column base plates may simply rest on the floor slab or they may be anchored to the floor slab. Racks are often joined back-to-back so they can be loaded and unloaded from aisles on both sides.

The stored products rest on pallets, which are supported by shelves. The most common type of shelf is made from epoxy-coated stainless-steel wires; it is known as the wire-mesh shelf. The coefficient of friction between the pallets and the wire-mesh shelves is low—between 0.11 and 0.29. Therefore, the inertial forces transmitted by the pallets to the rack are also low. In the CA direction, the rack is very stiff due to the presence of trusses. Therefore, the rack deforms less but the pallets slide more. Excessive sliding can cause the pallets to fall off the shelves. In the DA direction, the rack is very flexible due to the presence of partially restrained moment frames. Therefore, the rack deforms more but the pallets slide less. Excessive swaying of rack in the DA direction can cause the beam–column moment connections to fail and the rack to collapse.

The seismic response of storage racks is quite nonlinear even at low levels of ground shaking. The nonlinearities are due to: (1) sliding of pallets on shelves; (2) large rotations in beam–column moment connections; and (3) moments induced by the large deflections of the gravity loads (P–Δ effect). The seismic performance of racks under design level ground shaking cannot be assessed without a nonlinear analysis.

Only a handful of experimental and analytical studies [1–5] have been conducted on the seismic response of storage racks. Focused mainly on the response of racks in the DA direction, the past analytical studies did not model sliding of pallets on shelves.

The objectives of this chapter are to:

- Present nonlinear-static and nonlinear-dynamic analyses of storage racks;
- Identify factors which control the seismic performance of racks; and
- Gain insight into the seismic response of racks so that sound decisions can be made to improve their performance during future earthquakes.

7.2   Structural System

Figure 7.1 shows the side and front views of a one-bay storage rack. The side view (left side) shows the vertical truss which resists the lateral load in the CA direction. The vertical truss occurs at the ends of each bay. A rack with n bays has $n + 1$ vertical trusses. The front view (right side) shows the moment frame which resists the lateral load in the DA direction. There are two moment frames (front and back) of each bay.

The rack analyzed in this chapter is a one-bay, stand-alone structure with two brace frames in the CA direction and two moment frames in the DA direction. The rack has four shelves at 0.24 m, 1.94 m, 3.56 m, and 5.08 m above the base slab. Each shelf supports two loaded pallets weighing a total of 17.8 kN. The weight of the rack itself is only 2 kN. The lower bound value of the dynamic (kinematic) coefficient of friction between the pallets and the shelves is $\mu = 0.11$. The upper bound value of the static coefficient of friction between the pallets and the shelves is $\mu = 0.29$.

The proprietary beam–column moment connections can be described as partially restrained connections. They differ from the fully restrained moment connections, discussed in Chap. 3 and 4, in following ways:

- A partially restrained moment connection is weaker than the connecting members. The plastic rotations occur within the connection itself. A fully restrained moment connection is stronger than the connecting members. The plastic rotations occur in the connecting members—usually beams.

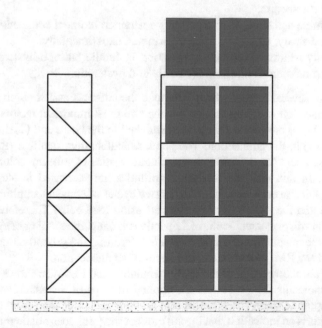

Fig. 7.1 Lateral load resisting system of a storage rack in the CA direction (left) and DA direction (right). For the sake of clarity, loaded pallets are only shown in the DA direction

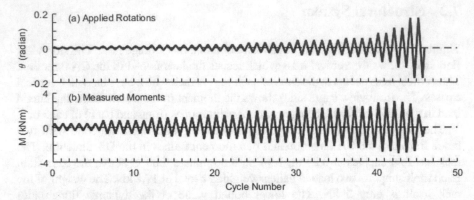

Fig. 7.2 Cyclic moment–rotation test: (**a**) applied rotations of various amplitudes and (**b**) moments mobilized in the connection

- In a partially restrained moment connection, the energy loss due to hysteresis is quite small. In a fully restrained moment connection, plastic yielding in beam is a major source of energy loss. This source of energy loss is absent from a partially restrained moment connection.
- A partially restrained moment connection is much more flexible compared to a fully restrained moment connection.
- Under cyclic loading, the strength and stiffness of a partially restrained moment connection degrade much more rapidly compared to those for a fully restrained moment connection.
- The stiffness and strength of a partially restrained moment connection cannot be calculated easily; they have to be determined experimentally.
- A partially restrained moment connection is ductile but non-hysteretic. A fully restrained moment connection is ductile and hysteretic.

There are several factors which influence the stiffness and strength of a beam–column connection in storage racks, but the two most important factors are: (1) the length (height) of the connector element attached to the beam and (2) the number of bolts (or pins) in the connection. The Rack Manufacturers Institute (RMI) [6] has standardized a test for determining the characteristics of a beam–column moment connection. In this test, the rotation amplitudes are increased in steps until the connection fails. In each step, six, four, or two cycles of constant amplitude rotations are applied and the moments are measured using load cells. Therefore, the test is performed in "displacement–control." Strictly speaking, the number of cycles should depend on the magnitude of the earthquake controlling the ground motion hazard at the site, but the RMI test [6] lacks that level of sophistication.

Figure 7.2a shows a plot of the applied rotations and Fig. 7.2b shows a plot of the measured moments. The rotations are applied in steps of increasing amplitude. In smaller amplitude steps, the moment amplitude does not change from one cycle to the next. This is an indication that the stiffness of the joint does not degrade from one cycle to the next. In larger amplitude steps, the moment amplitude reduces

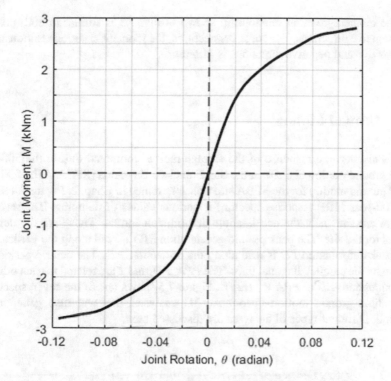

Fig. 7.3 Moment–rotation relationship of the connection from the results of cyclic test shown in Fig. 7.2

significantly from one cycle to the next. This is an indication that the stiffness of the joint degrades from one cycle to the next. As the rotations become larger and larger, the degradation in stiffness becomes more and more significant. In the last step, the degradation is so significant that the joint breaks even before completing the desired number of cycles for that step. The test is stopped at this stage.

From the results of the RMI test shown in Fig. 7.2, a cyclic moment–rotation relationship for the connection is generated as follows. From Fig. 7.2a, the rotation amplitude is read for each step. The corresponding moment amplitude is read from Fig. 7.2b as the smallest amplitude for that step, which usually occurs in the last cycle of the step. Figure 7.3 shows the resulting moment–rotation relationship for the connection. Note the following:

- The slope of the line joining the origin with a point on the curve is known as the secant stiffness. The secant stiffness reduces with increase in rotation. The connection is nonlinear even at very small rotations.
- The connection fails at a rotation of 0.115 radian.
- The moment–rotation relationship is assumed nonlinear elastic. In other words, the hysteretic energy loss in the connection is neglected.

The moment–rotation relationship shown in Fig. 7.3 is similar in both positive and negative directions. For some connections, the moment–rotation relationship in the positive and negative directions is different.

7.3   Ground Motion

In this chapter, the response of the storage rack is computed during the 500-year MRP ground motion at a site in the San Francisco Bay Area (SFBA). The 500-year MRP ground motion for the SFBA site was determined in Chap. 2. Figure 7.4 shows the 500-year MRP response spectra for various values of damping. The response spectra are shown in the acceleration–deformation format. These are the demand curves for the site. The peak pseudo-acceleration *PPA* is read along the vertical axis and peak deformation *PD* is read along the horizontal axis. The natural period *T* is shown by the parallel diagonal lines. The *PPA*, *PD*, and *T* are related to each other by the expression: $PD = PPA \times (T/2\pi)^2$. Figure 7.5 shows one of the seven spectrum-compatible ground motion history for 2D analyses; this was also generated in Chap. 2. Different types of analyses are discussed next.

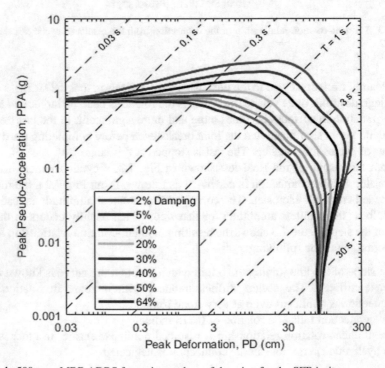

Fig. 7.4 500-year MRP ADRS for various values of damping for the SFBA site

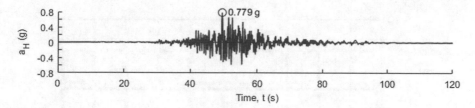

Fig. 7.5 One of seven spectrum-compatible history of 500-year MRP horizontal ground motion for the SFBA site

7.4 Cross-Aisle Responses

In the CA direction, the lateral loads are resisted by the vertical trusses (Fig. 7.1). For low levels of shaking, the pallets remain bonded with the shelves and the system behaves in a linear fashion with a natural period T_{CA} and damping of about 2% of critical. During linear response, each pallet experiences the same acceleration as the supporting shelf. When the acceleration of the shelf exceeds μg, the pallet starts sliding on the shelf. During sliding, the acceleration of the pallet remains fixed at μg but the acceleration of the shelf is greater than μg. Since the coefficient of friction is usually small, the frictional forces transmitted by the sliding pallets are not very high hence the deformation of the truss is not significant. The most important response in the CA direction is the amount of sliding. If sliding is too high, the pallets could fall off the shelves. Two methods of calculating the sliding response are discussed: (1) nonlinear static and (2) nonlinear dynamic.

7.4.1 Nonlinear-Static Analysis in CA Direction

The nonlinear-static analysis is carried out by generating a capacity curve and a damping curve for the storage rack. It is assumed that: (1) the weight of the storage rack is negligible compared with the weight of the pallets; (2) all pallets slide in unison with each; and (3) the coefficient of friction between pallets and shelves is $\mu = 0.11$ (lower bound estimate). With these assumptions, a pushover curve for the storage rack is generated; it is shown in Fig. 7.6. Initially, the force increases linearly with deformation. The slope of the force–deformation relationship is the lateral stiffness of the storage rack in the CA direction:

$$k_{CA} = m\left(\frac{2\pi}{T_{CA}}\right)^2 \tag{7.1}$$

where $m = 907 \times 8 = 7260$ kg = total mass of the loaded pallets; $T_{CA} = 0.33$ s = natural period of the rack-pallet system before sliding. Sliding begins

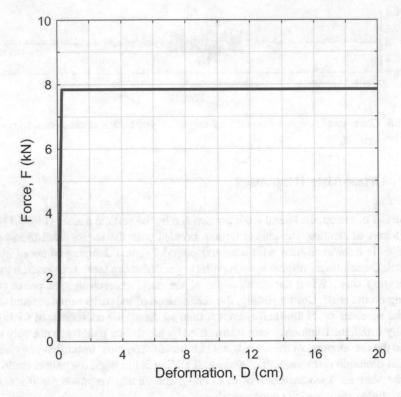

Fig. 7.6 Pushover curve for the storage rack in the CA direction

when the lateral force reaches $F = \mu mg = 0.11 \times 7260 \times 9.81 = 7.83$ kN. During sliding, the lateral force remains fixed at 7.83 kN.

The capacity curve is generated by dividing the lateral force F by the mass $m = 7260$ kg to obtain acceleration. Figure 7.7 shows a plot of the capacity curve. Figure 7.8 shows the same capacity curve on a logarithmic scale. Note that: (1) the period of the system before sliding is $T_{CA} = 0.33$ s, (2) the maximum acceleration experienced by the pallets is 0.11 g, and (3) the rack deforms 0.29 cm before any sliding occurs. During sliding, the pallet acceleration remains fixed at 0.11 g and the rack deformation remains fixed at 0.29 cm.

The damping curve is generated from cyclic force deformation relationships of various amplitudes. Figure 7.9 shows the cyclic force deformation relationship for 10-cm amplitude. The hysteretic energy per cycle E_h is the area of the loop on the left side. The "strain energy" E_s is the area of the triangle on the right side. For a 10-cm cycle, $E_h = 3$ kNm and $E_s = 0.39$ kNm. The equivalent-viscous hysteretic damping is given by the following expression [7]:

$$\zeta_h = \frac{E_h}{4\pi E_s} \tag{7.2}$$

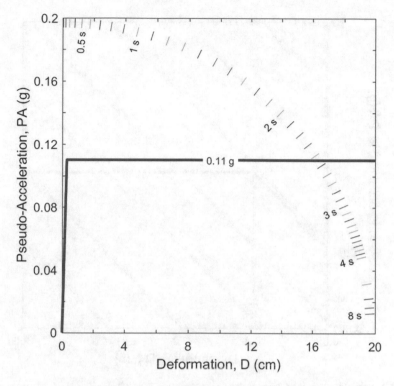

Fig. 7.7 Capacity curve generated from the pushover curve in Fig. 7.6

For a 10-cm cycle, the hysteretic $\zeta_h = 0.61$ (61% of critical). The hysteretic damping is similarly computed for cycles of other amplitudes. Figure 7.10 shows a plot of hysteretic damping versus amplitude. This is the raw damping curve, which needs to be adjusted as follows:

1. During seismic shaking, the rack will experience cycles of many different amplitudes. The smaller amplitude cycles will have lower damping than larger amplitude cycles. Since hysteretic damping increases with an increase in amplitude, the damping for the entire duration of shaking should be smaller than the damping for the peak amplitude cycle. The average damping for 10-cm deformation is computed as follows. The area under the raw damping curve up to 10-cm deformation is 5.51 cm (Fig. 7.11). Dividing this area by 10 cm gives an average damping of 0.55 (55% of critical). The average damping is similarly computed for other values of deformation.

2. Finally, the damping is not allowed to drop below 2% of critical to account for other sources of damping besides pallet sliding.

The plot in Fig. 7.12 shows the adjusted damping curve which will be used in the nonlinear-static analysis.

In Fig. 7.13, the capacity curve (Fig. 7.8) is superimposed on the 500-year MRP response spectra (demand curves) for 2, 5, 10, 20, 30, 40, 50, and 64% damping. The

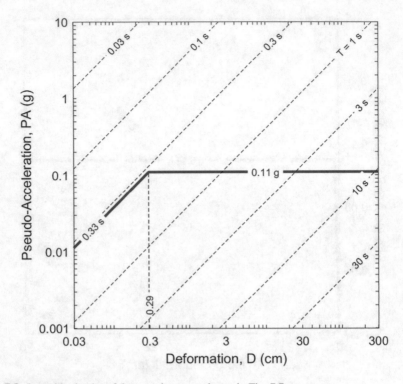

Fig. 7.8 Logarithmic plot of the capacity curve shown in Fig. 7.7

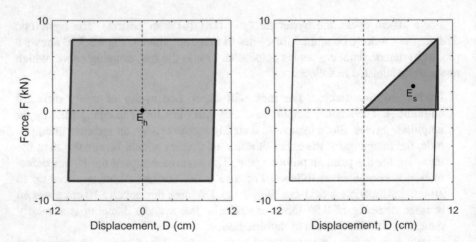

Fig. 7.9 Hysteretic and "strain" energies for 10-cm amplitude cycle

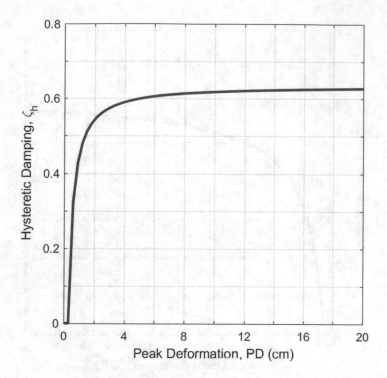

Fig. 7.10 Hysteretic damping for cycles of various amplitudes

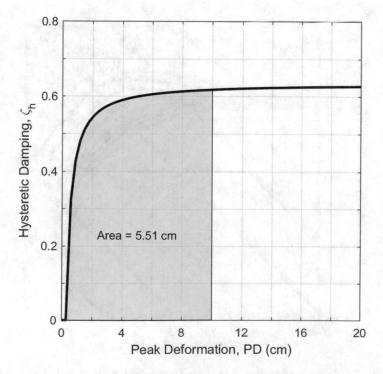

Area = 5.51 cm

Fig. 7.11 Computing adjusted hysteretic damping

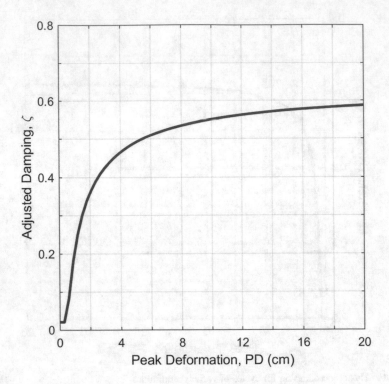

Fig. 7.12 Adjusted damping-versus-deformation curve

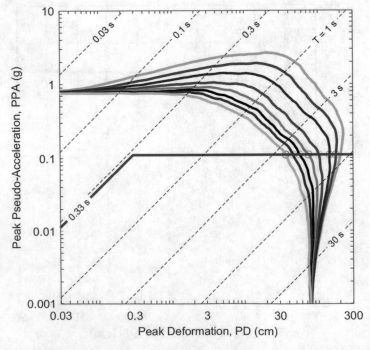

Fig. 7.13 Capacity curve superimposed on demand curves for 2, 5, 10, 20, 30, 40, 50, and 64% of critical damping

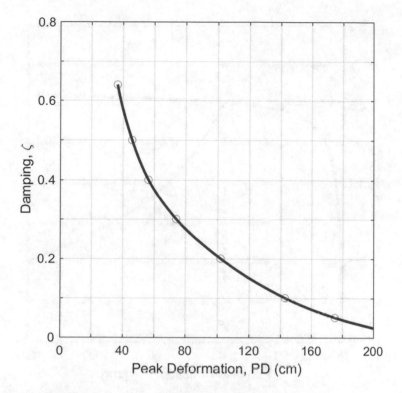

Fig. 7.14 Deformation-versus-damping curve for the rack in the CA direction

intersections of capacity curve with demand curves represent equilibrium points for various assumed values of damping. Figure 7.14 shows deformations for various assumed values of damping. If damping is high, the deformation is low. This is known as the deformation-versus-damping curve, even though the deformation is shown along the horizontal axis. The deformations are for various assumed values of damping, but the actual value of damping is not known yet.

There are two relationships between damping and deformation. Figure 7.14 shows that the deformation reduces with increase is damping. Figure 7.12 shows that the damping increases with increase in deformation. Both of these relationships have to be satisfied. In Fig. 7.15, the deformation-versus-damping curve (Fig. 7.14) is superimposed on the damping-versus-deformation curve (Fig. 7.12). These two curves intersect at a damping of 0.61 (61% of critical) and deformation of 38 cm. Subtracting elastic deformation of 0.29 cm from 38 cm gives sliding displacement of 37.7 cm. The shelves are 1.04 m wide. A 37.7 cm shift in the CG will not cause the pallets to slide off the shelves, but they will be hanging precariously.

To estimate the lateral forces in the rack, an upper bound value of the coefficient of friction is used, which, in this case, is $\mu = 0.29$. The frictional force on the rack is $0.29 \times 17.8 = 5.16$ kN at each level. Although pallets cannot experience an acceleration greater than 0.29 g, the rack itself, weighing 2 kN, can experience a

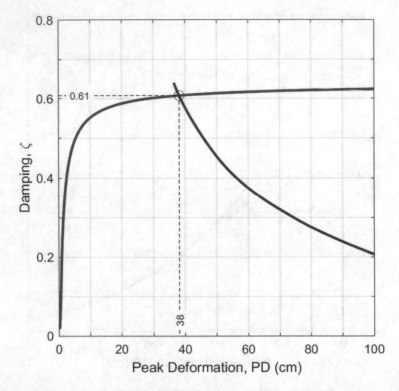

Fig. 7.15 Deformation-versus-damping curve superimposed on the damping curve

higher acceleration. The highest pseudo-acceleration on a 2% damping response spectrum is 2.65 g (Fig. 7.13). Multiplying this by the weight of the rack gives a base shear due to the inertia of the rack itself. The total base shear for the rack is $Q = 5.16 \times 4 + 2 \times 2.65 = 26$ kN and the total overturning moment for the rack is $OTM = 70$ kNm; these are shared by the two vertical trusses.

The results are next confirmed by performing a nonlinear-dynamic analysis.

7.4.2 Nonlinear-Dynamic Analysis in CA Direction

Figure 7.16 shows a computer model of the rack created in SAP 2000 [8]. The columns and braces are modeled by *linear beams*. The pallet masses are lumped at each shelf level. Elastoplastic *link elements* [8] are introduced between the pallet masses and beams to simulate friction. The hysteretic characteristics of the *link elements* are based on Ref. [9]. The lateral strength of the *link element* is 0.11 times the pallet weight, thus simulating a friction coefficient of $\mu = 0.11$. The deformation of the *link element* represents sliding of pallet on shelf.

Fig. 7.16 Computer model
of the rack [8] in CA
direction

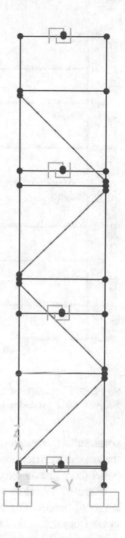

The model of the rack in the CA direction is excited at its base by the first simulation of the 500-year MRP ground motion shown in Fig. 7.5. As long as the shelf acceleration is less than 0.11 g, the pallet moves with the shelf. When the shelf acceleration exceeds 0.11 g, the pallet starts sliding on the shelf and its acceleration remains fixed at 0.11 g. Figure 7.17 displays pallet sliding at each level. The maximum sliding is 32 cm and it occurs at the first level. The sliding displacements are approximately the same at every level. This implies that the rack is moving like a "rigid" structure in the CA direction. Since the pallet forces cannot exceed frictional resistance and the rack is quite stiff in the CA direction (due to the presence of trusses), it does not deform much.

To estimate the forces in the rack, an upper bound estimate of the friction coefficient is used, which in this case is $\mu = 0.29$. Figure 7.18 displays the lateral forces in the *link elements* at various levels. These are the pallet forces (pallet mass ×

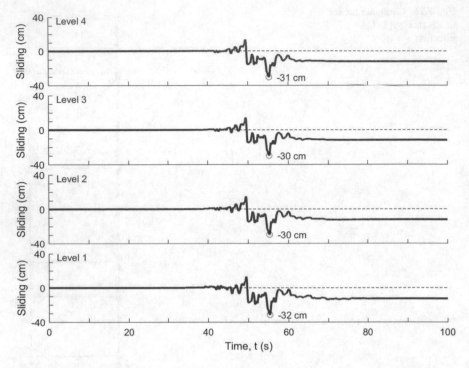

Fig. 7.17 Cross-aisle pallet sliding at various levels of the rack subjected to the first simulation of 500-year MRP ground motion generated in Chap. 2

acceleration) exerted on the rack at various levels. The pallet forces cannot exceed the maximum frictional resistance, which is $0.29 \times 17.8 = 5.16$ kN. Therefore, the pallet forces never exceed 5.16 kN and the pallet acceleration never exceeds 0.29 g.

Figure 7.19a shows a plot of the base shear, which controls the axial force in the diagonal braces. The maximum base shear is 28.2 kN. This is 8% higher than the value from the static analysis. The base shear due to pallet forces at all four levels is $5.16 \times 4 = 20.6$ kN. The remaining 7.6 kN base shear is due to the acceleration of the rack itself. Since the racks weigh 2 kN, the effective acceleration of the rack is $7.6/2 = 3.8$ g. Figure 7.19b shows a plot of the overturning base moment, which controls the: (1) axial force in columns; (2) axial force in anchors; and (3) vertical force on the base slab. The maximum overturning moment is 79 kNm. This is 13% higher than the value from the static analysis. The overturning moment due to pallet forces at four levels is $5.16 \times (0.24 + 1.94 + 3.56 + 5.08) = 55.8$ kNm. The remaining 23.2 kNm overturning moment is due to the acceleration of the rack itself.

The results of pallet sliding, base shear, and overturning moment were obtained for seven simulated ground motions. A summary of results is presented in various rows of Table 7.1. The median values in the last row of Table 7.1 are the responses to the 500-year MRP ground motion at the site.

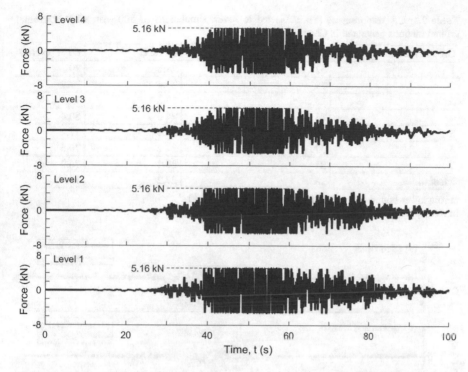

Fig. 7.18 Frictional forces applied on the rack by pallets at various levels during the first simulation of 500-year MRP ground motion generated in Chap. 2

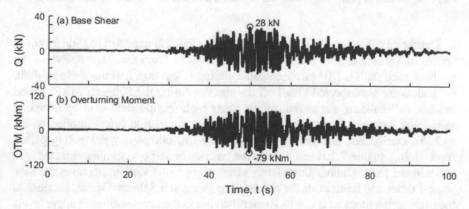

Fig. 7.19 CA base-shear and overturning base moment for rack subjected to the first simulation of 500-year MRP ground motion generated in Chap. 2

Table 7.1 CA responses of rack subjected to seven simulations of 500-year MRP horizontal ground motions generated in Chap. 2

Ground Motion	Pallet Sliding[a] (cm)	Q[b] (kN)	OTM[b] (kNm)
1	32	28.2	79
2	29	26.4	77.2
3	28	27.6	79.6
4	23	27.8	80.4
5	20	26.4	77.4
6	37	27.2	79.6
7	21	26.4	76.2
Median	**28**	**27.2**	**79**

aFrom lower bound estimate of friction coefficient $\mu = 0.11$
bFrom upper bound estimate of friction coefficient $\mu = 0.29$

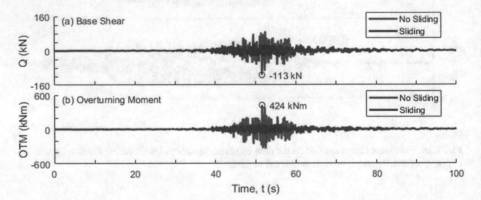

Fig. 7.20 Effects of preventing sliding on: (**a**) base shear; and (**b**) overturning base moment

During the 500-year MRP ground motion, the pallet is expected to slide 28 cm as shown in Table 7.1. This is 25% less than the value from the static analysis. The width of each shelf is 1.04 m. Since pallet sliding is less than half the width of shelf, the pallets are not expected to fall off the shelves during the 500-year MRP ground motion. Pallet sliding can be reduced by using high-friction shelves or by installing restraining bars, but those measures will increase the inertial forces applied on the rack. To clarify that, new results were generated by completely preventing pallets from sliding. Figure 7.20a and b show the base shear and overturning moment with and without pallet sliding. Completely eliminating pallet sliding increases the base shear 4 times and it increases the overturning moment 5.5 times. Greater increase in the overturning moment is due to greater increase in the accelerations at upper levels when sliding is prevented. An alternative approach to reduce pallet sliding is by making the rack more flexible in the transverse direction, say by eliminating some of the diagonal braces. By letting some deformations to occur in the rack, the amount of sliding can be reduced.

In summary,

- Friction between pallets and shelves limits the inertial forces transmitted from pallets to the rack.
- Increasing the friction between pallets and shelves reduces pallet sliding but it increases the forces experienced by the rack and the base slab.
- Limited sliding of pallets is beneficial to the seismic performance of racks. Sliding should be allowed as long as it does not pose a problem.
- Sliding displacement can be reduced by making the rack more flexible in the cross-aisle direction.

7.5 Down-Aisle Responses

In the down-aisle (DA) direction, the storage rack is much more flexible, hence it is not expected to experience high enough accelerations to induce sliding of pallets. Even if the accelerations are high enough to induce sliding, there is not enough room between the pallets and the columns to allow significant unobstructed sliding. Therefore, in most cases, it is reasonable to assume that the pallets will not slide in the DA direction. There are two important sources of nonlinearity that need to be modeled in the DA direction—joint rotations and P–Δ effect. As already discussed, the moment–rotation relationship for beam-column joints is nonlinear. Also, large deflections increase the moments in the joints by offsetting the gravity loads. Once again, the analysis is carried out in two different ways: (1) nonlinear static and (2) nonlinear dynamic.

7.5.1 Nonlinear-Static Analysis in DA Direction

The hysteretic action in the partially restrained beam-column connections is small but it is not negligible. Therefore, 5% of critical damping is considered reasonable in the DA direction. There is no need to develop the damping curve, but the rest of the analysis is similar to that for the multistory moment frame discussed in Chap. 4.

In generating the pushover curve, the columns and beams are assumed "rigid" compared to the joints. The CG of all masses is at height H from the base. It is deflected horizontally by an amount D. From geometry, the rotation in each of the 16 joints is:

$$\theta = \frac{D}{H} \tag{7.3}$$

The moment $M(\theta)$ is read from Fig. 7.3 as a function of the joint rotation θ. From static equilibrium, the lateral force F is given by the following expression:

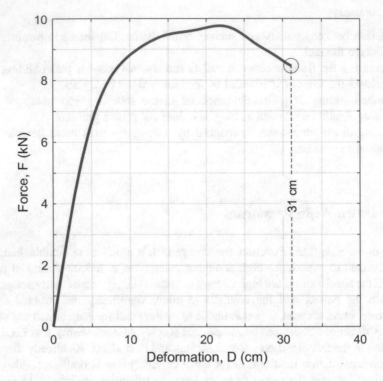

Fig. 7.21 Pushover curve for the rack in the DA direction

$$F = \frac{16M(\theta)}{H} - \frac{D \cdot W}{H} \qquad (7.4)$$

where W = weight of the fully loaded rack. The second term on the right side of
Eq. 7.4 is due to the P–Δ effect. Figure 7.21 shows a plot of F versus D. This is the
pushover curve for the storage rack in the DA direction. Initially, the force
F increases with deflection D because the first term on the right-hand side of
Eq. 7.4 is more dominant. At larger deflections, the second term become dominant
and the force starts decreasing with increase in deflection. The pushover curve stops
at deflection $D = 31$ cm when the rotation θ reaches the limiting value of 0.115 radian
for the joints.

The capacity curve is obtained by dividing the force F by the weight W.
Figure 7.22 shows the capacity curve for the rack in the DA direction. Capacity
curve is a plot of pseudo-acceleration versus deflection. The radial tick marks
indicate the effective period of the rack. The maximum value of pseudo-acceleration
is 0.14 g, which is insufficient to slide the pallets assuming an upper bound value of
the coefficient of friction $\mu = 0.29$. Figure 7.23 shows a logarithmic plot of the
capacity curve shown in Fig. 7.22. The radial tick marks in Fig. 7.22 are replaced by
parallel diagonal lines in Fig. 7.23.

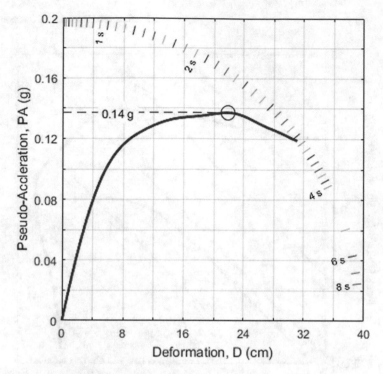

Fig. 7.22 Capacity curve for the rack in the DA direction

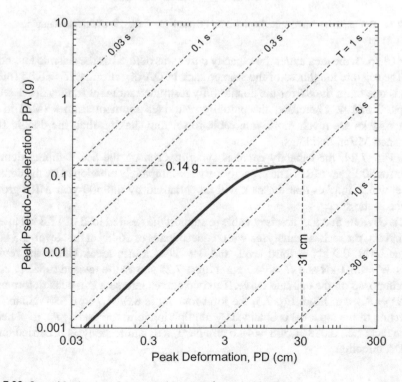

Fig. 7.23 Logarithmic plot of the capacity curve shown in Fig. 7.22

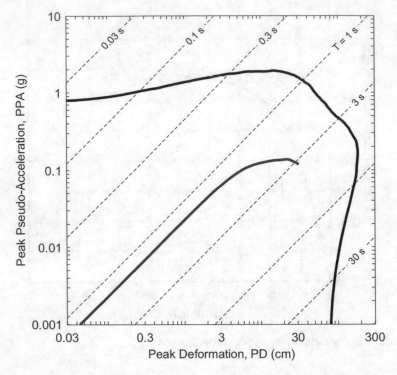

Fig. 7.24 DA capacity curve superimposed on demand curve for 5% of critical damping

In Chap. 3, the area under the capacity curve was defined as the seismic toughness *ST*. The seismic toughness of the storage rack in DA direction is $ST = 0.35$ (m/s)^2. This is one-tenth the *ST* for the ductile fully restrained moment frames discussed in Chaps. 3 and 4. Therefore, the partially restrained moment frames (MF) in the storage racks are much more vulnerable to seismic shaking than the ductile fully restrained MF in buildings.

In Fig. 7.24, the capacity curve is superimposed on the 5% damping demand curve for 500-year MRP. The capacity curve is completely enveloped by the demand curve, which implies that the rack will be collapsed by the 500-year MRP ground motion at the site.

It is obvious that the rack is not able to support the desired load of 17.8 kN at each level. Next, the rack is reanalyzed with reduced loads of 16 kN at first level, 5.3 kN at second level, 0.9 kN at third level, and 0.9 kN at fourth level. Under the revised loads, $W = 23.1$ kN and $H = 94.7$ cm. Figure 7.25 shows the revised capacity curve superimposed on the demand curve. The two curves intersect at a peak deformation of $PD = 8.3$ cm. From Eq. 7.3, the joint rotation is 8.3/94.7 = 0.088 radian. The deflection of the top level is obtained by multiplying joint rotation by the total height of the rack; i.e., $0.088 \times 508 = 44.6$ cm. Next, a dynamic analysis is carried out in the DA direction.

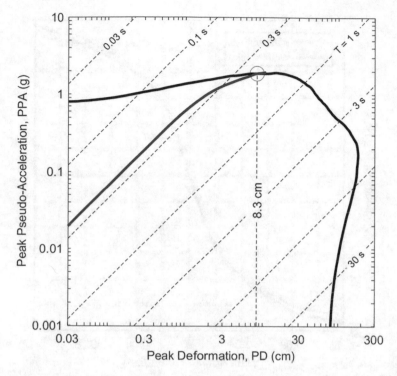

Fig. 7.25 DA capacity curve of lightly loaded rack superimposed on demand curve for 5% of critical damping

7.5.2 Nonlinear-Dynamic Analysis in DA Direction

As in the case of static analysis, two sources of nonlinearity need to be modeled in the dynamic analysis. These are: (1) nonlinear rotations in the joints and (2) P–Δ effect. The P–Δ effect induces additional moment in the joints; therefore, it effectively softens the joints. To understand this, the Eq. 7.4 can be rewritten as follows:

$$F = \frac{16(M(\theta) - \theta HW/16)}{H} \tag{7.5}$$

For each value of θ, $\theta HW/16$ is subtracted from the joint moment $M(\theta)$ to model the softening of the joint due to P–Δ. Figure 7.26 shows the moment–rotation relationship for the softened joint to account for P–Δ effect.

Figure 7.27 shows a computer model of the rack in the DA direction. The beams and columns are modcled as *linear beams* in SAP 2000 [8]. The joints are modeled as nonlinear rotational springs with the moment–rotation relationship shown by the

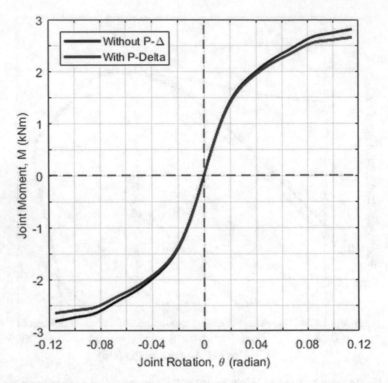

Fig. 7.26 Softening of the joint due to P–Δ effect

blue curve in Fig. 7.26. The model is subjected to the first simulation of 500-year MRP ground motion shown in Fig. 7.5. Figure 7.28 shows histories of deflections at various levels. The maximum deflection is 38 cm at the top level. Figure 7.29 shows histories of joint rotations at various levels. The maximum rotation is 0.071 radian at the third level. The rack was analyzed for all seven simulations of 500-year MRP ground motion. The results are listed in Table 7.2. The median values, in the last row of Table 7.2, are the responses to the 500-year MRP ground motion. The deflection from dynamic analysis is 15% less than that from the static analysis. The rotation from dynamic analysis is 26% less than that from the static analysis.

Large deflections of the rack can cause it to impact adjacent structures or walls. Large deflections can also damage any brittle pipes attached with the rack. The rack sway and joint rotations can be reduced by using stiffer and stronger beam–column connections. As mentioned before, the stiffness and strength of beam–column connections are most influenced by the length (height) of the beam-end connector element and by the number of bolts (pins) in the connection.

Fig. 7.27 Computer model of the rack in down-aisle direction

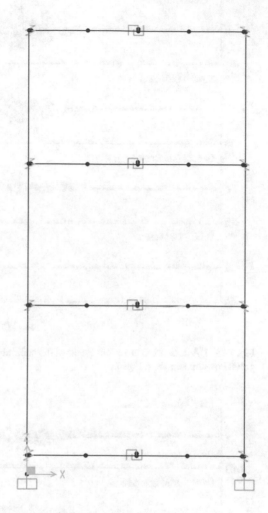

7.6 Summary

1. The mass of stored product on a rack is several times the mass of the rack itself. But the product is not rigidly attached to the rack.
2. Sliding of the product on the shelves limits the inertial forces transmitted to the rack.
3. In the CA direction, the rack is very stiff. Therefore, the rack sways less, but the product slides more.
4. In the DA direction, the rack is very flexible. Therefore, the rack sways more, but the product slides less.
5. The CA response of racks is controlled by the coefficient of friction between pallets and shelves.

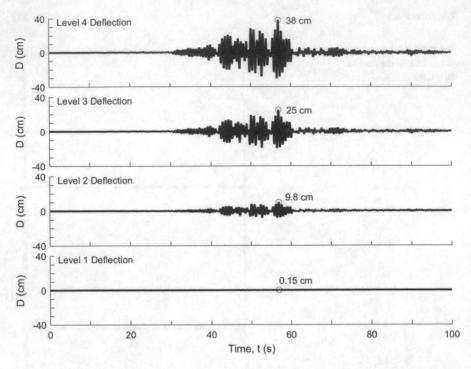

Fig. 7.28 DA deflections at various levels of the rack subjected to first simulation of 500-year MRP ground motion shown in Fig. 7.5

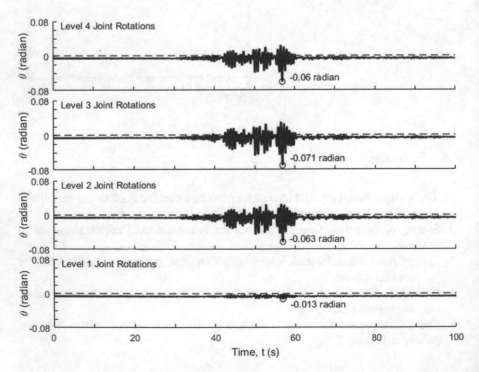

Fig. 7.29 Hinge rotations in beam-column moment connections at various levels of the rack subjected to first simulation of 500-year MRP ground motion shown in Fig. 7.5

Table 7.2 Responses of rack in down-aisle direction due to seven simulated horizontal ground motions generated in Chap. 2

Ground motion	Top sway (cm)	Joint rotation (radian)
1	38.2	0.0707
2	35.5	0.0653
3	31.7	0.0563
4	35	0.0608
5	39	0.0717
6	34.6	0.0627
7	35.3	0.0649
Median	**35.3**	**0.0649**

6. The DA response of racks is controlled by the strength and deformability of beam–column moment connections.
7. Sliding in the CA direction can be reduced by replacing standard shelves with high-friction shelves. However, increased friction between product and shelves increases the forces applied to the rack.
8. As an alternative, the product sliding in the CA direction can be reduced by making the rack more flexible by allowing some bending to occur in the columns.
9. Sway and joint rotations in the DA direction can be reduced by increasing the stiffness and strength of moment connections.
10. In general, the seismic toughness of storage racks is much less than that of building frames. Therefore, racks are much more vulnerable to ground motions than building frames.

References

1. Chen, C. K., Scholl, R. E., & Blume, J. A. (1980a). Seismic study of industrial storage racks. In *Report prepared for the National Science Foundation and for the rack manufacturers institute and automated storage and retrieval systems (sections of the material handling institute)* (p. 569). San Francisco, CA: URS/John A. Blume & Associates.
2. Chen, C. K., Scholl, R. E., & Blume, J. A. (1980b). Earthquake simulation tests of industrial steel storage racks. In *Proceedings, 7th world conference on earthquake engineering* (pp. 379–386). Istanbul.
3. Filiatrault, A., & Wanitkorkul, A. (2004). Shake table testing of Frazier industrial storage racks. In *Report No. CSEE-SEESL-2005-02, structural engineering and earthquake simulation laboratory, departmental of civil, structural and environmental engineering* (p. 83). State University of New York at Buffalo.
4. Bernuzzi, C., Chesi, C., Parisi, M. A. (2004). *Seismic Behavior and Design of Steel Storage Racks*. In 13th World Conference on Earthquake Engineering Vancouver, B.C., Canada, August 1-6, 2004, Paper No. 481.
5. Filiatrault, A., Bachman, R. E., & Mahoney, M. G. (2006). Performance-based seismic Design of Pallet-Type Steel Storage Racks. *Earthquake Spectra, 22*(1), 47–64.

6. Rack Manufacturers Institute (RMI). (2002). "Specification for the design, testing, and utilization of industrial steel storage racks." 2002 Edition, Charlotte, NC.
7. Chopra, A. K. (2011). *Dynamics of Structures,* 4th edition, Prentice-Hall International Series in Civil Engineering and Engineering Mechanics
8. Computers and Structures. (2011). *CSI analysis reference manual for SAP 2000.* Berkeley, CA: Computers and Structures Inc.
9. Wen, Y. K. (1976). Method for random vibration of hysteretic systems. *Journal of the Engineering Mechanics Division, ASCE, 102,* EM2.

Chapter 8
Seismic Response of Liquid-Storage Tanks

Nomenclature

ζ_h	Equivalent-viscous hysteretic damping
μ	Coefficient of friction between steel base and foundation
g	Acceleration due to gravity (9.81 m/s^2)
h_f	Depth of foundation (distance between top of soil to bottom of tank)
h_i	Impulsive liquid height for moment due to pressures on tank wall
h_i'	Impulsive liquid height for moment due to pressures on tank wall and base
m_c	Convective liquid mass
m_i	Impulsive liquid mass
m_t	Mass of tank, roof, base, and anchorage
PPA_c	Convective response acceleration
PPA_i	Impulsive response acceleration
SSI	Soil-structure interaction
T	Natural period
T_e	Effective period

8.1 Introduction

Figure 8.1 shows the vertical section of a water storage steel tank resting on a concrete mat foundation. During horizontal shaking of the ground, the liquid near the bottom of the tank moves with the tank shell, while the liquid near the top of the tank sloshes [1–6]. The former liquid is known as the "impulsive liquid" and the latter liquid is known as the "convective liquid." The aspect ratio (height/radius) of the tank determines the relative proportions of impulsive and convective liquids. For slender tanks, liquid is more constrained by the shell of the tank; therefore, the impulsive liquid is more. For broad tanks, liquid is less constrained by the shell of

© Springer Nature Switzerland AG 2021
P. K. Malhotra, *Seismic Analysis of Structures and Equipment*,
https://doi.org/10.1007/978-3-030-57858-9_8

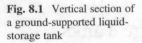

Fig. 8.1 Vertical section of a ground-supported liquid-storage tank

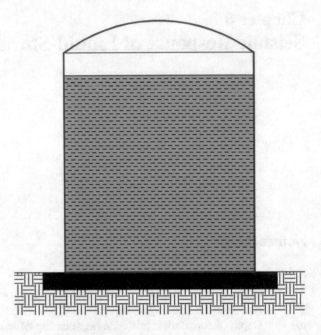

the tank; therefore, the convective liquid is more. For fixed-roof tanks with insufficient freeboard, the convective liquid is not able to slosh freely. Therefore, some of the convective liquid becomes impulsive [7].

The impulsive and convective liquids exert pressures on the tank shell and the tank base. The net effect of shell pressures is a shear and an overturning moment at the base of the tank shell. The shear is resisted by friction between the tank base and the foundation. The overturning moment is resisted by anchors which tie the tank shell to the foundation. The pressures on the tank base generate an additional overturning moment which is transmitted directly to the foundation. Vertical shaking of the ground generates axisymmetric pressures on the tank wall and the tank base [5]. Pressures on the wall generate hoop stresses. Pressures on the tank base result in a vertical force which is transmitted to the foundation.

The important sources of deformability and damping in a tank are:

1. **Soil deformation.** The flexibility of the soil below the foundation increases the deformability of the system and allows some vibration energy to radiate away from the tank.
2. **Base sliding.** Sliding of the tank on top of the foundation increases the deformability of the system and allows a significant loss of energy through friction.
3. **Base uplifting.** Overturning moment at the base of the tank shell can cause the anchors to yield and a portion of the tank to uplift on one side. Limited base uplifting can be a significant source of deformability for the tank. Uplifting is accompanied by yielding at the plate–shell junction, which can be a slight source of damping.

4. **Shell deformability.** The flexibility of the tank shell is a source of some deformability and small amount of damping.

The tank analyzed in this chapter is discussed next.

8.2 Example Tank

The tank in Fig. 8.1 is fabricated from stainless steel plates. It has a radius of $R = 6.8$ m and shell height of 15.6 m. It is filled with water to a height of $H = 14$ m. The freeboard in the tank is $15.6-14 = 1.6$ m. With a height-to-radius ratio of $14/6.8 = 2.1$, the tank is considered slender. The maximum height of fixed dome roof on top of shell is 2.3 m. The tank shell is anchored to the concrete mat with 30 stainless steel straps, each of cross-sectional area 12.9 cm². The thickness of the mat foundation is $h_f = 1.22$ m and its radius is $R_f = 8.4$ m. Table 8.1 lists the

Table 8.1 Tank data

Radius	6.8 m		
Shell height	15.6 m		
Water height	14 m		
Freeboard	1.6 m		
Roof dome height above shell	2.3 m		
Tank material (shell, base plate, dome)	A240–304 SS		
Shell courses	Course	Width (m)	Thickness (mm)
	6	2.28	4.76
	5	2.42	4.76
	4	2.42	4.76
	3	3.03	6.35
	2	3.03	6.35
	1	2.42	12.71
Base plate thickness	6.35 mm		
Mass of tank (shell, base plate, and roof)	47,200 kg		
Coefficient of friction at the base of tank	0.57		
Anchor straps	Number	30	
	Cross-sectional area	19.4 cm²	
	Length	171 cm	
	Material	A240–304 SS	
Foundation mat	Radius	8.4 m	
	Thickness	1.22 m	
Allowable soil bearing pressure	144 kN/m²		
Ultimate soil bearing pressure	431 kN/m²		
"Zero-strain" shear modulus of the soil, G_0	144 MN/m²		

important properties of the tank and foundation. The mass of the tank is $m_t = 47.2 \times 10^3$ kg. The mass of liquid in the tank is $m_l = 2.03 \times 10^6$ kg.

8.3   Ground Motion

In this chapter, the response of the storage tank is computed during the 500-year MRP ground motion at a site in the San Francisco Bay Area (SFBA). The 500-year MRP ground motion for the SFBA site was determined in Chap. 2. Figure 8.2 shows the 500-year MRP response spectra for various values of damping. The response spectra are shown in the acceleration–deformation format. These are the demand curves for the site. The peak pseudo-acceleration *PPA* is read along the vertical axis and peak deformation *PD* is read along the horizontal axis. The natural period *T* is shown by the parallel diagonal lines. The *PPA*, *PD*, and *T* are related to each other by the expression: $PD = PPA \times (T/2\pi)^2$. Only static analyses are performed in this chapter. Therefore, ground motion histories are not needed.

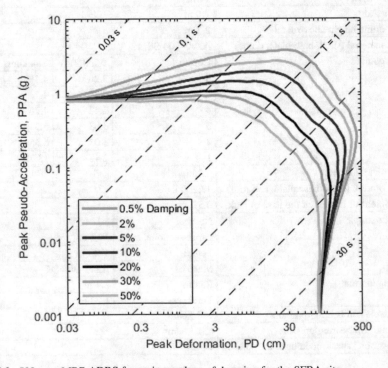

Fig. 8.2 500-year MRP ADRS for various values of damping for the SFBA site

8.4   Linear Analysis of Fixed-Base Tank

First, a linear analysis of the tank is performed by ignoring the flexibility of the soil below the foundation. This is the simplest and most common type of analysis for tanks.

8.4.1   Model of Fixed-Base Tank

Figure 8.3 shows a linear model of the fixed-base tank as per Ref. [6]. The liquid mass m_l is divided into two parts: (a) the impulsive liquid mass of 1.56×10^6 kg and (b) the convective liquid mass of $m_c = 0.47 \times 10^6$ kg. The mass of the tank is lumped with the impulsive mass to obtain the total impulsive mass of $m_i = 1.56 \times 10^6 + 47.2 \times 10^3 = 1.61 \times 10^6$ kg. The impulsive period is $T_i = 0.177$ s and the convective period is $T_c = 3.86$ s. The impulsive damping is $\zeta_i = 0.02$ (2% of critical) and the convective damping is $\zeta_c = 0.005$ (0.5% of critical).

So far, it is assumed that the liquid in the tank can slosh freely without touching the dome roof. In other words, the sloshing wave height is assumed to be less than the freeboard of 1.6 m. The validity of this assumption is checked next.

8.4.2   Sloshing Wave Height

The convective peak pseudo-acceleration, read from the 0.5% damping response spectrum in Fig. 8.4, is $PPA_c = 0.556$ g. The sloshing wave height is [6]:
  $d = (PPA_c/g) \times R = 3.78$ m.

Fig. 8.3 Linear model of the fixed-base tank [6]

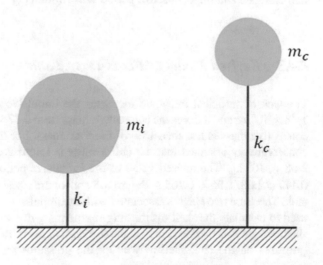

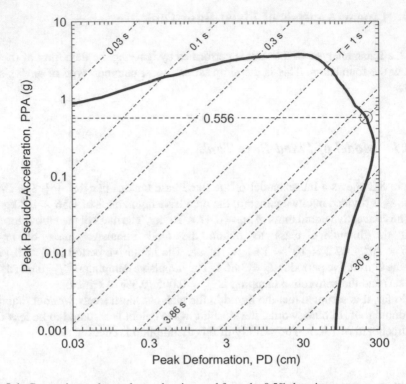

Fig. 8.4 Convective peak pseudo-acceleration read from the 0.5% damping response spectrum

But the freeboard in the tank is only 1.6 m (Table 8.1). Therefore, the sloshing liquid will touch the dome roof. The constraining action of the dome roof will transform some of the convective liquid to impulsive [7]. Also, the impulsive period will elongate and the convective period will shorten [7].

8.4.3 Revised Model of Fixed-Base Tank

The lack of sufficient freeboard increases the impulsive mass from 1.61×10^6 to 1.78×10^6 kg and reduces the convective mass from 0.47×10^6 to 0.3×10^6 kg. The convective mass is less than 15% of the total mass. For the sake of simplicity, it is conservatively assumed that the entire mass is impulsive for this tank, i.e., $m_i = 2.08 \times 10^6$ kg. The revised value of the impulsive period as per Ref. [7] is $T_i = 0.177\sqrt{2.08/1.56} = 0.205$ s. Figure 8.5 shows the revised model of the fixed-base tank. There are two heights associated with the impulsive mass: h_i and h_i'. Height h_i is used to calculate the shell overturning moment M_S due to the pressures on the tank shell. Height h_i' is used to calculate the foundation overturning moment M_F due to the pressures on the tank shell as well as the pressures on the tank base. The shell

Fig. 8.5   Revised linear
model of the fixed-base tank

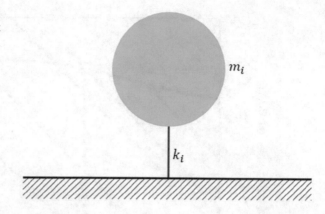

Table 8.2   Parameters of linear fixed-base model

Impulsive mass, m_i	2.08×10^6 kg
Impulsive period, T_i	0.205 s
Impulsive spring stiffness, k_i	1960 MN/m
Impulsive damping constant, c_i	2.55 MNs/m
Height of impulsive mass for calculating shell moment, h_i	6.28 m
Height of impulsive mass for calculating foundation moment, h_i'	6.97 m

overturning moment M_S is used to calculate the stresses in the tank shell and anchors.
The foundation overturning moment M_F is used to calculate the soil pressures and
the stresses in the mat foundation. Some additional parameters, for use in later
analyses, are:

Impulsive spring stiffness:

$$k_i = m_i \left(\frac{2\pi}{T_i}\right)^2 = 1,960 \text{ MN/m}$$

and impulsive damping constant:

$$c_i = 2\zeta_i m_i \left(\tfrac{2\pi}{T_i}\right) = 2.55 \text{ MNs/m}$$

Table 8.2 lists important parameters of the linear fixed-base model.

8.4.4   Responses of Fixed-Base Tank

In Fig. 8.6, the impulsive pseudo-acceleration is read from the 2% damping response
spectrum, corresponding to the impulsive period $T_i = 0.205$ s; it is $PPA_i = 2.02$ g.
The base shear is obtained by multiplying the impulsive mass m_i with the impulsive
pseudo-acceleration PPA_i, i.e., $Q = 2.08 \times 10^6 \times 2.02 \times 9.81 = 41.2$ MN. The shell
overturning moment is obtained by multiplying the base shear by height h_i, i.e., M_S

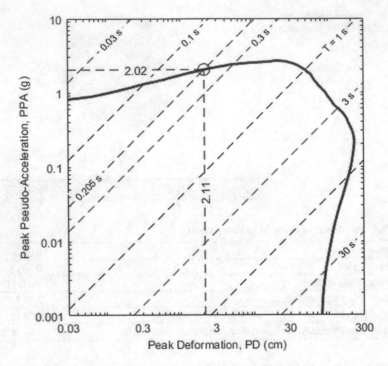

Fig. 8.6 Impulsive response acceleration for fixed-base tank

Table 8.3 Responses from linear analyses		Rigidly supported	Flexibly supported
	PPA_i	2.02 g	1.36 g
	PD_i	2.11 cm	4.15 cm
	Q	41.2 MN	27.6 MN
	M_S	259 MNm	173 MNm
	M_F	337 MNm	226 MNm

$= 41.2 \times 6.28 = 259$ MNm. The foundation overturning moment is obtained by multiplying the base shear by the height $\left(h_i' + h_f\right)$, i.e. $M_F = 41.2 \times (6.97 + 1.22) = 337$ MNm. The peak deflection of the impulsive mass is $PD_i = PPA_i(T_i/2\pi)^2 = 2.11$ cm. The values of PPA_i, PD_i, Q, M_S, and M_F are listed in Table 8.3. This concludes the linear-static analysis of the fixed-base tank.

The maximum frictional resistance at the base of the tank is $0.577 \times (47.2 \times 10^3 + 2.03 \times 10^6) \times 9.81/10^6 = 11.8$ MN. The base shear is 3.5 times the maximum frictional resistance. The maximum moment capacity of anchor straps is 64.8 MNm. The shell overturning moment is four times the moment capacity of anchor straps. The maximum overturning capacity of the foundation is 139 MNm. The foundation overturning moment is 2.4 times the moment capacity of the foundation. In other words, the strength demand on the tank is very high based on the linear analysis of fixed-base tank. Next, the linear model of the tank is refined to include some additional sources of deformability and damping.

8.4.5   Linear Analysis of Flexibly Supported Tank

The flexibility of the soil below the foundation is an additional source of deformability; it causes the impulsive period to elongate. Soil flexibility also allows some vibration energy to radiate away from the tank. Therefore, in most cases, soil flexibility increases the damping of the system. The effects of soil flexibility on the impulsive period and damping are captured by the soil–structure interaction (SSI) analysis discussed in this section.

8.4.5.1   Model of Flexibly Supported Tank

Figure 8.7 shows the model used in the SSI analysis. The horizontal and the rotational soil springs in the SSI model (Fig. 8.7) simulate the lateral and rocking flexibility of the foundation soil. The horizontal and rotational spring constants k_x and k_θ depend on the size of the foundation and the shear modulus of soil; they are determined as per Ref. [8] which is based on Refs. [9, 10]. k_x and k_θ are listed in Table 8.4. The horizontal and the rotational dampers c_x and c_θ simulate the loss of energy through radiation. For the sake of clarity, horizontal and rotational dampers are not shown in Fig. 8.7, but they are in parallel with the horizontal and rotational

Fig. 8.7 Linear model of flexibly supported tank

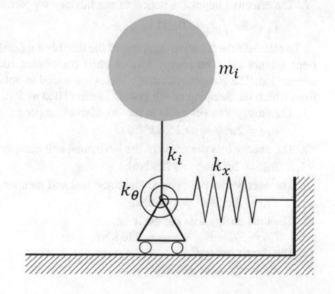

Horizontal spring stiffness, k_x	2390 MN/m
Rotational spring stiffness, k_θ	111,780 MNm/radian
Horizontal damping constant, c_x	66.5 MNs/m
Rotational damping constant, c_θ	1080 MNms/radian

springs. The damping constants also depend on the size of the foundation and the shear modulus of soil; they are determined as per Ref [8] and are listed in Table 8.4.

Next, the period and damping of flexibly supported tank are calculated. A value of PPA_i is assumed and the deformation in each component (spring) of the model is calculated. The component deformations are added to determine the total deformation of the system which is then used to calculate the period of the system. The procedure is illustrated below:

1. Assume $PPA_i = 1$ g $= 9.81$ m/s^2. The base shear is:
   $Q = m_i PPA_i = 2.08 \times 10^6 \times 9.81 = 20.38$ MN
2. The deformation of the impulsive spring is:
   $u_i = Q/k_i = 20.38/1960 = 0.0104$ m
3. The deformation of the horizontal soil spring is:
   $u_x = Q/k_x = 20.38/2390 = 0.0085$ m
4. The moment in the rotational soil spring is:
   $M_F = Q \cdot \left(h_i' + h_f\right) = 20.38 \times (6.97 + 1.22) = 166.8$ MNm
5. The rotation of the soil spring is:
   $\theta = M_F/k_\theta = 166.8/111,780 = 0.00149$ radian
6. The total deflection of the impulsive mass is:
   $u_t = u_i + u_x + \theta(h_i + h_f) = 0.0104 + 0.0085 + 0.00149 \times (6.28 + 1.22)$
   $= 0.0301$ m
7. The effective impulsive period of the flexibly supported tank is:

$$T_e = 2\pi\sqrt{\frac{u_t}{PPA_i}} = 0.351 \text{ s}.$$

To calculate the effective damping of the flexibly supported tank, some additional steps are needed. The energy loss in each component (damper) of the model is determined. The component energy losses are added to obtain the total energy loss, from which the damping of the system is calculated as follows:

8. The energy loss per cycle in the impulsive damper is:
   $E_{Di} = 2\pi^2 \frac{c_i}{T_e} u_i^2 = 15.6$ kNm
9. The energy loss per cycle in the horizontal soil damper is:
   $E_{Dx} = 2\pi^2 \frac{c_x}{T_e} u_x^2 = 272$ kNm
10. The energy loss per cycle in the rotational soil damper is:
    $E_{D\theta} = 2\pi^2 \frac{c_\theta}{T_e} \theta^2 = 141$ kNm
11. The total energy loss per cycle is:
    $E_D = E_{Di} + E_{Dx} + E_{D\theta} = 428$ kNm
12. The "strain energy" is:
    $E_S = \frac{1}{2} Q \cdot u_t = 312$ kNm

13. The effective damping for the flexibly supported tank is:
    $\zeta_e = \frac{E_D}{4\pi E_S} = 0.11$ (11% of critical).

Since the model is linear, the period and damping do not depend on the assumed value of PPA_i.

8.4.6 Responses of Flexibly Supported Tank

The impulsive period and damping of the flexibly supported tank are 0.351 s and 11% of critical, respectively. The 11% damping response spectrum is generated by interpolating between the 10 and 20% damping response spectra for the site (Fig. 8.2). The 11% damping response spectrum is shown in Fig. 8.8. The impulsive pseudo-acceleration and deformation, read from the 11% damping response spectrum, are $PPA_i = 1.36$ g and $PD_i = 4.15$ cm (Fig. 8.8). Table 8.3 compares the responses of the flexibly supported tank with those of the rigidly supported tank. The strength demands on the flexibly supported tank are smaller than those on the rigidly supported tank because the former has higher deformability and damping than the latter.

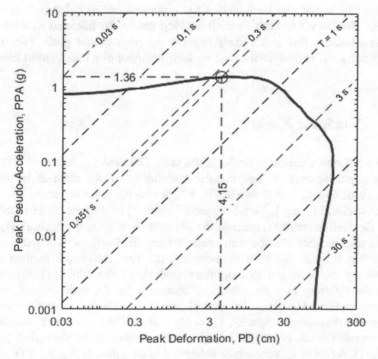

Fig. 8.8 Impulsive response acceleration and deformation for flexibly supported tank

Strength demands on the tank are still very high. The base shear of 27.6 MN is still high enough to slide the tank. The shell overturning moment of 173 MNm is still high enough to yield the anchors and uplift the tank on its foundation. The foundation overturning moment of 226 MNm is still high enough to yield the soil on one side and uplift the foundation on the other side. In Chaps. 3–7, yielding, sliding and uplifting were shown to be important sources of deformability and damping which can reduce the strength demands on a structure. Limited amounts of yielding, sliding, and uplifting can be tolerated as long as they do not prevent the tank from meeting its performance objectives. Next, a nonlinear analysis of the tank is performed to consider these additional sources of deformability and damping.

8.5   Nonlinear Analysis of Tank

Nonlinear analysis makes it possible to consider all sources of deformability and damping which can reduce the strength demand on the tank. In a nonlinear analysis, the anchors can yield and the tank uplift on one side. The tank can also slide if the horizontal shear overcomes friction between the tank base and the foundation. The foundation of the tank can lift off the soil on one side, and the soil below the foundation can yield on the opposite side. It is important to note that only those nonlinearities which can be tolerated (or accepted) are modeled in a nonlinear analysis. Nonlinearities such as shell-bucking cannot be tolerated in most cases because a buckled tank is not likely to meet any performance goals. Therefore, it is not worthwhile to analyze the post-bucking behavior of a tank in most cases.

8.5.1   Nonlinear Model

Figure 8.9 shows a nonlinear model of the tank. The model is described below.

The impulsive mass m_i and impulsive spring k_i are the same as before. The impulsive damper C_i (not shown in Fig. 8.9) is also the same as before.

The rotational spring k_ϕ simulates partial uplifting of the tank on its foundation. When the shell moment M_S exceeds the ultimate capacity of the anchors M_{anc}, the anchors start yielding and the tank starts rotating (rocking) on its foundation. The relationship between the shell moment and the base rotation is nonlinear; it is obtained by performing a base-uplifting analysis of the tank [11]. Figure 8.10 shows the uplifting moment–rotation relationship for the tank without anchors. The base moment M_B along the vertical axis is the difference between the shell moment and the anchors' capacity, $M_B = M_S - M_{anc}$. The area of the hysteresis loop E_ϕ represents the loss of energy through plastic yielding at the plate–shell junction (Fig. 8.11). A plot of E_ϕ versus base rotation ϕ is presented in Fig. 8.12; this is later used in the generation of a damping curve for the tank.

Fig. 8.9 Nonlinear model of the tank

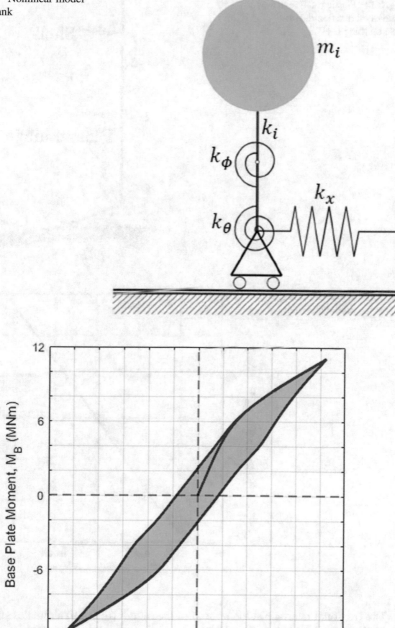

Fig. 8.10 Moment–rotation relationship at the base of unanchored tank

Fig. 8.11 Plastic rotation at
plate–shell junction due to
base uplifting [11]

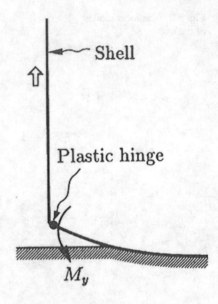

Fig. 8.12 Hysteretic energy
loss at the base plate–shell
junction

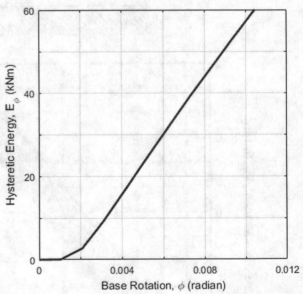

The frictional interface at the bottom of the model (Fig. 8.9) simulates sliding of
the tank on its foundation. The frictional interface prevents the impulsive pseudo-
acceleration from exceeding a limiting value PPA_{slide} which is determined by the
coefficient of friction between the tank and the foundation. Even though the frictional interface is shown at the bottom of the model in Fig. 8.9, it simulates sliding of
the tank on top of the foundation.

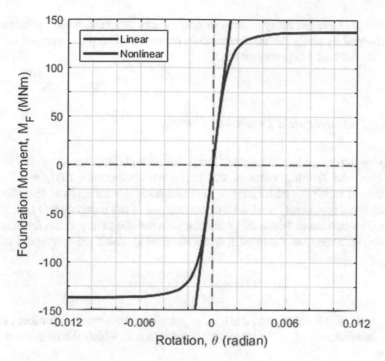

Fig. 8.13 Moment–rotation relationship for the mat foundation

The horizontal soil spring k_x, the horizontal soil damper C_x and the rotational soil damper C_θ (not shown in Fig. 8.9) are same as before. The rotational soil spring is no longer linear. It is extended into the nonlinear range to simulate foundation lift off and soil yielding. The following steps are taken to obtain the nonlinear soil spring:

1. The foundation is assumed to rest on a bed of compression-only springs of collective vertical stiffness $k_\theta \times 4/R_f^2$, where k_θ is the linear stiffness of the rotational soil spring, and $R_f =$ the radius of the foundation.
2. The springs are modeled as elastic–plastic; they remain linear up to the bearing strength of the soil and then yield without mobilizing any additional force.
3. The springs are pre-compressed by the total weight of the tank plus the foundation.
4. The rotation of the foundation is increased in small steps and the corresponding moment is calculated in each step from analysis.

Figure 8.13 shows a plot between the foundation rotation θ and the foundation overturning moment M_F. The red line in Figure 8.13 corresponds to the linear stiffness k_θ of the rotational soil spring. As the foundation uplifts and/or the soil below the foundation starts yielding, the rotational soil spring becomes nonlinear. The ultimate overturning strength of the foundation is $M_{ult} = 139$ MNm.

Although, there are many sources of nonlinearity in a tank, a stable analysis can be performed by using the nonlinear-static procedure. The pushover and damping curves for the tank are generated next.

8.5.2   Pushover and Damping Curves

A step-by-step procedure to generate the pushover and damping curves is described here. First, the limiting value of the impulsive acceleration $PPA_i = PPA_{lim}$ is calculated at which the tank slides or the foundation overturns. Note that sliding of the tank is not necessarily a failure but overturning of the foundation is catastrophic failure. The maximum frictional resistance between the tank and the foundation is $\mu(m_i + m_c)g$. Therefore, the impulsive acceleration at which the tank slides is:

$$PPA_{slide} = \frac{\mu(m_i + m_c)g}{m_i}$$

Since $m_c = 0$ for this tank, $PPA_{slide} = \mu g$. Since the ultimate moment capacity of the foundation is M_{ult}, the impulsive acceleration at which the foundation overturns is:

$$PPA_{overturn} = \frac{M_{ult}}{m_i(h_i' + h_f)}$$

The limiting value of the peak pseudo-acceleration is smaller of PPA_{slide} and $PPA_{overturn}$:

$$PPA_{limit} = min\left(PPA_{slide}, PPA_{overturn}\right)$$

The impulsive pseudo-acceleration PPA_i is increased in small steps from 0 to PPA_{lim}. For each value of PPA_i, the base shear, deflection, and damping are calculated as follows:

1. The base shear is:

$$Q = m_i PPA_i$$

2. The deformation of the impulsive spring is:

$$u_i = \frac{Q}{k_i}$$

3. The shell overturning moment is:

$$M_S = Qh_i$$

4. The shell overturning moment is compared with the ultimate capacity of the anchors M_{anc}. If $M_S \leq M_{anc}$, the base rotation $\phi = 0$. If $M_S > M_{anc}$, base rotation ϕ is read from Fig. 8.10 corresponding to the base moment $M_B = M_S - M_{anc}$. M_B is the moment in the rotational spring k_ϕ in Fig. 8.9. The hysteretic energy loss per cycle E_ϕ is read from Fig. 8.12.

5. The horizontal deflection of the foundation is:

$$u_x = \frac{Q}{k_x}$$

6. The foundation overturning moment is:

$$M_F = Q(h_i' + h_f)$$

7. M_F is the moment in rotational spring k_θ in Fig. 8.9. From Fig. 8.13, the rotation θ of the foundation is read against M_F. The elastic value of the foundation rotation, used later in damping calculation, is:

$$\theta_{el} = \frac{M_F}{k_\theta}$$

8. The total deflection of the impulsive mass is:

$$u_t = u_i + u_x + \phi h_i + \theta(h_i + h_f)$$

9. The effective impulsive period of the tank is:

$$T_e = 2\pi \sqrt{\frac{u_t}{PPA_i}}$$

10. The energy dissipated per cycle is:

$$E_D = 2\pi^2 \frac{c_i}{T_e} u_i^2 + 2\pi^2 \frac{c_x}{T_e} u_x^2 + 2\pi^2 \frac{c_\theta}{T_e} \theta_{el}^2 + E_\phi$$

11. The "strain energy" is:

$$E_S = \frac{1}{2} Q_i u_t$$

12. The effective damping ratio is:

$$\zeta_e = \frac{E_D}{4\pi E_S}$$

During sliding

The pushover and damping curves during sliding are computed by gradually increasing the sliding displacement u_s while keeping the impulsive pseudo-acceleration fixed at $PPA_i = PPA_{slide}$. The u_x, u_i, θ, and ϕ are kept fixed at their last computed values for $PPA_i = PPA_{limit}$. The following additional steps are needed to continue the pushover and damping curves during sliding:

13. The total deflection of the impulsive mass is

$$u_t = u_i + u_x + \phi h_i + \theta(h_i + h_f) + u_s$$

14. The effective impulsive period during sliding is

$$T_e = 2\pi \sqrt{\frac{u_t}{PPA_{slide}}}$$

15. The overall energy dissipated per cycle is

$$E_D = 2\pi^2 \frac{c_i}{T_e} u_i^2 + 2\pi^2 \frac{c_x}{T_e} u_x^2 + 2\pi^2 \frac{c_\theta}{T_e} \theta_{el}^2 + E_\phi + 4\mu(m_i + m_c)g u_s$$

The first 4 terms on the right-hand side of the above expression are the same as those on the right-hand side of expression under Step 10. The last term is the energy dissipated by friction.
16. The "strain energy" E_S is computed from the expression under Step 11 and the effective damping ratio ζ_e is computed from the expression under Step 12.

Figure 8.14 shows a plot between the total deflection u_t and the base shear Q. This is the pushover curve for the tank. It is insightful to look at Fig. 8.15 which provides relative contributions of various sources of deformation. For very small amounts of total deformation, nearly 65% of the deformation occurs in the foundation soil and 35% occurs in the tank shell. Therefore, it is obvious that the foundation flexibility is important even at very low levels of shaking. At 1.6 cm deformation, the anchors yield and the tank starts uplifting on one side. From that point onward, the contributions of the foundation and shell deformations start reducing dramatically. At 7.88 cm deformation, the tank slides on its foundation and sliding becomes the main source of deformation for the tank.

Figure 8.16 shows a plot between the total deflection u_t and the impulsive peak pseudo-acceleration PPA_i. This is the capacity curve. Figure 8.17 shows the capacity curve on a logarithmic scale.

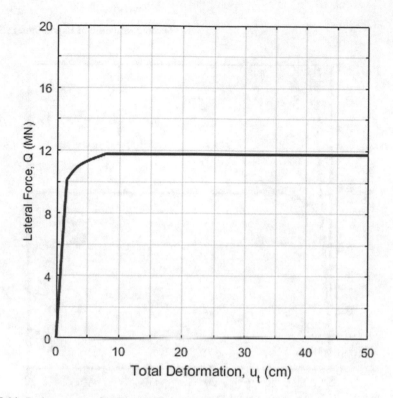

Fig. 8.14 Pushover curve for the tank

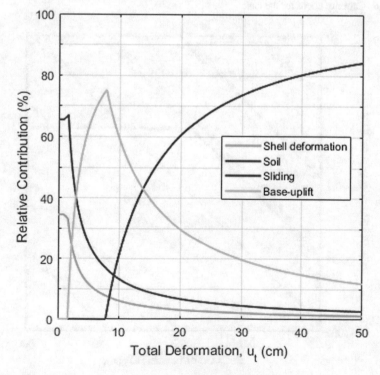

Fig. 8.15 Relative contributions of various sources of deformation

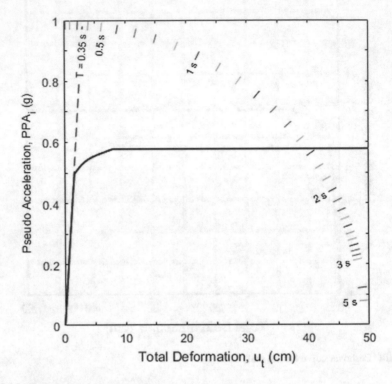

Fig. 8.16 Capacity curve for the tank

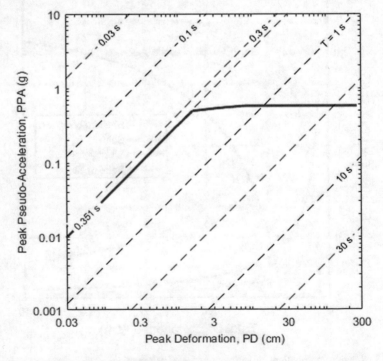

Fig. 8.17 Logarithmic plot of the capacity curve

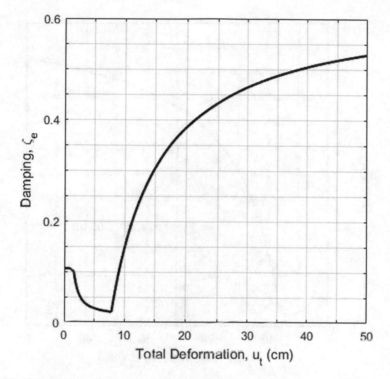

Fig. 8.18 Damping for various values of deformation

Figure 8.18 shows a plot between the total deflection u_t and the effective damping ζ_e. This is the "raw" damping curve for the tank. Two most important sources of damping are: (1) radiation of vibration energy through the foundation soil and (2) sliding of the tank. In the beginning, the damping is dominated by radiation. The radiation damping is due to terms with c_x and c_θ in steps 10 and 15 in Section 8.5.2. Damping reduces when the tank starts uplifting because damping associated with uplifting is small. Finally, when the tank starts sliding, damping increases dramatically. Damping is lowest in the region where the dominant source of deformation is base uplifting. The damping in this region can be increased by anchoring the tank with energy-dissipating anchors [12].

The damping curve of Fig. 8.18 is not yet ready for nonlinear-static analysis because during seismic response the peak deformation occurs only once; rest of the times the deformation is less than the peak. Since the damping for smaller amplitude cycles is different, the damping is adjusted as follows. For $u_t = 25$ cm, the damping is 0.43 according to Fig. 8.18. The average (adjusted) damping for deformations between 0 and 25 cm is calculated by taking the area under the curve in Fig. 8.19 up to 25 cm and dividing that area by 25 cm. This gives the adjusted damping of 5.55/25 = 0.22 (22% of critical). The adjusted damping is similarly computed for other values of total deformation. Figure 8.20 shows the adjusted damping curve for the system; this will be used to complete the nonlinear-static analysis.

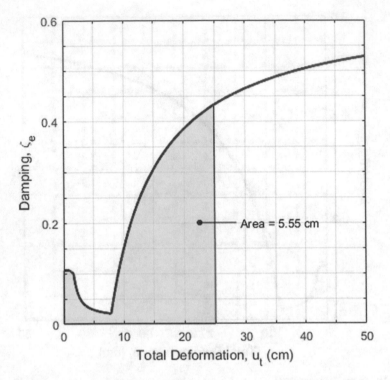

Fig. 8.19 Damping adjustment

8.5.3 Pushover Analysis

In Fig. 8.21, the capacity curve is superimposed on demand curves (response spectra) for various values of damping. The intersections of the capacity and the demand curves provide the deformations for various assumed values of damping. These are shown in Fig. 8.22. Since deformation depends on damping according to Fig. 8.22 and damping depends on deformation according to Fig. 8.20, an intersection between these two curves will provide the deformation and damping at equilibrium. These two curves intersect at a deformation of 22.9 cm and damping of 20% of critical, as shown in Fig. 8.23. These are the values of deformation and damping at equilibrium. Figure 8.24 shows the equilibrium point on the capacity curve. Since the equilibrium point is in the flat portion of the capacity curve, the tank slides at its base. Total deflection is 22.9 cm, while sliding begins at 7.88 cm. Therefore, the amount of sliding is 22.9–7.88 = 15 cm. The impulsive acceleration at equilibrium is 0.577 g.

The last column of Table 8.5 provides a summary of responses from the nonlinear analysis. The tank uplifts 13.7 cm. The anchor straps should be able to elongate 13.7 cm without breaking. Stainless steel straps can stretch 10% before any "necking" occurs. Therefore, the straps should be at least 1.37 m long to avoid breaking.

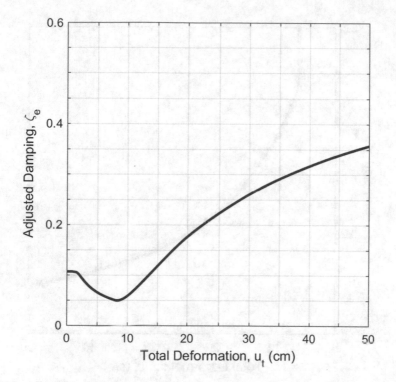

Fig. 8.20 Adjusted damping curve

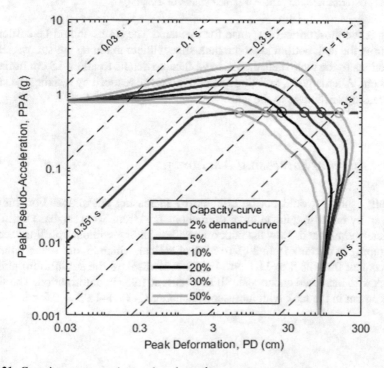

Fig. 8.21 Capacity curve superimposed on demand curves

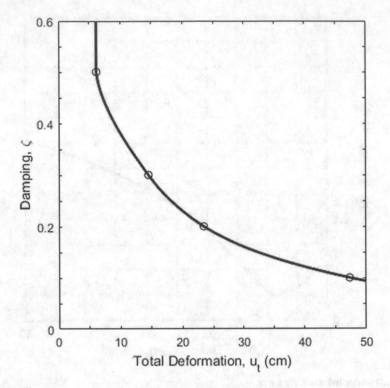

Fig. 8.22 Deformations for various assumed values of damping

Sliding is the most critical response for this tank. The straps should be sufficiently away from the tank bottom to let the tank slide without touching the straps. All pipe connections to the tank shell should have flexible details to allow 15 cm horizontal movement. Alternatively, uplifting and sliding can be reduced by resizing the tank or by increasing anchorage to the tank.

8.5.4   Effect of Increasing Anchorage

Normally, sliding would not be affected by increasing anchorage. For this tank, however, by increasing the number of anchor straps from 30 to 36, base uplifting is completely eliminated. Now the tank only slides. Sliding without uplifting increases the damping of the tank from 20% to 36% of critical, which in turn reduces the total displacement from 24.4 to 11.1 cm. Figure 8.25 displays the equilibrium point for the tank with increased anchorage. Sliding starts at 1.95 cm deformation. The sliding displacement in the tank with increased anchorage is 11.1–1.95 = 9.2 cm.

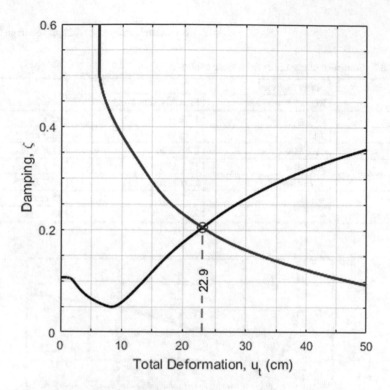

Fig. 8.23 Intersection of deformation-versus-damping curve of Fig. 8.22 with damping-versus-deformation curve of Fig. 8.20

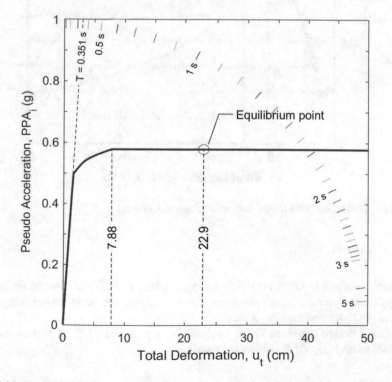

Fig. 8.24 Equilibrium point on the capacity curve

Table 8.5 Summary of responses from three analyses

	Linear analyses		Nonlinear analysis
	Rigidly supported	Flexibly supported	
PPA_i	2.02 g	1.36 g	0.577 g
PD_i	2.11 cm	4.15 cm	22.9 cm
Q	41.2 MN	27.6 MN	11.8 MN
M_S	259 MNm	173 MNm	88.3 MNm
M_F	337 MNm	226 MNm	96.7 MNm
Sliding	–	–	15 cm
Uplifting	–	–	13.7 cm

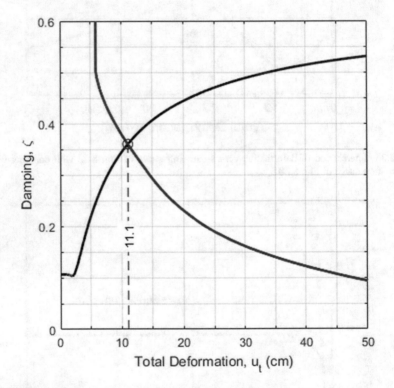

Fig. 8.25 Equilibrium point for the tank with increased anchorage

8.6 Summary

1. Linear analysis predicts very high strength demands on liquid-storage tanks.
2. Typically, strength demands from linear analysis are reduced by empirical strength reduction factors (R factors).
3. Seismic design based on linear analysis is very expensive (if R factors are not used) or arbitrary (if R factors are used).

4. There are some important sources of deformability and damping in tanks which can be utilized to reduce the strength demand on tanks.
5. A nonlinear analysis is needed to consider all sources of deformability and damping in tanks.

References

1. Jacobsen, L. S. (1949). Impulsive Hydrodynamics of Fluid inside a Cylindrical Tank and of Fluid Surrounding a Cylindrical Pier. *Bulletin of the Seismological Society of America, 39*(3), 189–203.
2. Housner, G. W. (1963). The dynamic behavior of water tanks. *Bulletin of the Seismological Society of America, 53*(2), 381–387.
3. Veletsos, A. S.; Yang, J. Y. (1977). "Earthquake response of liquid storage tanks." In *Proceedings of the Second Engineering Mechanics Specialty Conference, ASCE, Raleigh*, 11–24.
4. Haroun, M. A., & Housner, G. W. (1981). Seismic Design of Liquid-storage Tanks. *Journal of Technical Councils, 107*(1), 191–207.
5. Veletsos, A. S. (1984). "Seismic response and Design of Liquid Storage Tanks." In *American Society of Civil Engineers, New York, Guidelines for the Seismic Design of Oil and Gas Pipeline Systems, prepared by the Committee on Gas and Liquid Fuel Lifelines of the ASCE Technical Council on Lifeline Earthquake Engineering.* Chapter 7, 1984.
6. Malhotra, P., Wenk, T., & Wieland, M. (2000). Simple procedure for seismic analysis of liquid-storage tanks. *Journal of Structural Engineering International, 10*(3), 197–201.
7. Malhotra, P. K. (2006). Earthquake induced sloshing in cone and dome roof tanks with insufficient freeboard. *Journal of Structural Engineering International, 16*(3), 222–225.
8. ASCE. (1998). Seismic analysis of safety related nuclear structures and commentary. In *ASCE Standard No. ASCE 4–98.* Reston, VA: American Society of Civil Engineers.
9. Veletsos, A. S. (1977). Dynamics of structure foundation systems. In M. Newmark & W. J. Hall (Eds.), *Structural and Geotechnical Mechanics, a volume honoring N* (pp. 333–361). Englewood Cliffs, NJ: Prentice Hall.
10. Gazetas, G. (1991). Formulas and charts for impedances of surface and embedded foundations. *ASCE Journal of Geotechnical Engineering, 117*, 9.
11. Malhotra, P. K. (1997). Seismic response of soil-supported unanchored liquid-storage tanks. *Journal of Structural Engineering, 122*(4), 440–450.
12. Malhotra, P. K. (2000). Practical nonlinear seismic analysis of tanks. *Earthquake Spectra, 16*(2), 473–492.

Chapter 9
Seismic Response of Gantry Cranes

Nomenclature

ζ	Damping ratio (fraction of critical)
ζ_h	Hysteretic damping ratio (fraction of critical)
CP	Collapse Prevention limit state
E_h	Hysteretic energy loss per cycle
E_s	"Strain energy"
IO	Immediate Occupancy limit state
MRP	Mean return period
PD	Peak deformation (also known as spectral deformation; sometimes called spectral displacement)
PGA	Peak ground acceleration
PPA	Peak pseudo-acceleration (also known as spectral acceleration)
s	Seconds
ST	Seismic toughness (area under the capacity curve)
T	Natural period of the structure (seconds)

9.1 Introduction

Gantry cranes are common in manufacturing facilities. Like most structures, they should remain operational during frequent earthquakes and not collapse during rare earthquakes. Figure 9.1 shows the sketch of a gantry crane. The gantry, weighing 587 kN (60,000 kg), moves on rails supported by runway beams on the north and south sides. There are two wheels on each side of the gantry. Differential movement between the north and south side wheels can cause the gantry crane to rotate about the vertical (z) axis. This rotation can cause the wheels (on north and south sides) to come closer and jam the gantry. To prevent jamming of the gantry due to rotation,

© Springer Nature Switzerland AG 2021
P. K. Malhotra, *Seismic Analysis of Structures and Equipment*,
https://doi.org/10.1007/978-3-030-57858-9_9

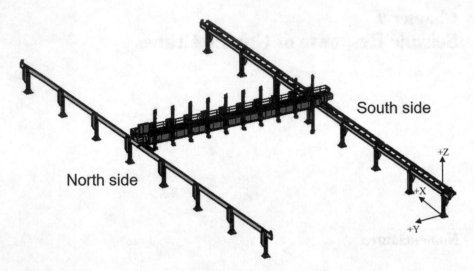

Fig. 9.1 Sketch of a gantry crane at a manufacturing facility

the south side wheels are allowed to move out freely up to 7.62 cm. The gantry beams are supported by 8 square-tube steel columns on each side (Fig. 9.1). The weight of the support structure (beams, columns, and rails) is 1.11 MN (113,000 kg). The total weight of the gantry and the support structure is 1.70 MN.

At their base, the columns are anchored to a 2.44 m wide, 1.22 m thick continuous concrete foundation. The center of gravity (CG) of the gantry is 5.08 m from the base. There are 61-cm high vertical stiffeners at the base of square-tube steel columns. Gantry is analyzed only in the transverse (north–south) direction. Under seismic loading, there are two critical configurations of the gantry crane:

1. **Configuration 1.** Gantry crane is centered between two columns. This configuration is critical for the runway beam.
2. **Configuration 2.** Gantry crane is centered over a column. This configuration is critical for the column.

Figure 9.2 shows a sketch of the gantry in Configuration 1. The distance of gantry wheel from the column is $a = 3.35$ m, distance between wheels is $b = 5.5$ m, and distance between columns is $L = 12.2$ m. In Fig. 9.2, P1 and P2 mark the centerlines of wheels, R1 and R2 mark the centerlines of columns.

Tables 9.1 and 9.2 list the important properties of the gantry columns and runway beams. Since the rail is attached on the top flange of the runway beams, only the top flange of the beams is assumed to resist the lateral seismic loads. The top flange of the beam and the square tube steel column cross section are determined to be "compact" according to AISC 360 [1]. Therefore, the entire sections can yield before buckling.

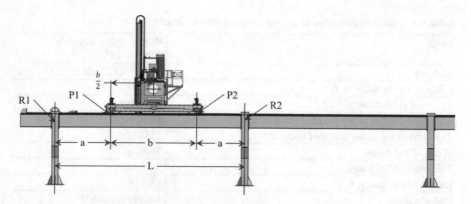

Fig. 9.2 Side view of the gantry in Configuration 1

Table 9.1 Important properties of the square tube steel columns

Form	HSS 18″ × 18″ × 5/8″
Material	ASTM A500 GR B
Minimum yield strength, f_y	317 MPa
Expected yield strength, $1.1f_y$	349 MPa
Width	45.7 cm
Depth	45.7 cm
Wall thickness	1.6 cm
Cross-section area	267 cm^2
Moment of inertia about the x-axis, I_x	84,210 cm^4
Height, H	5.08 m
Height of base stiffeners	61 cm
Young's modulus of elasticity, E	200 GPa
Elastic stiffness of column along y-direction, $3EI_x/H^3$	3.85 MN/m
Plastic modulus, S_x	4330 cm^3
Minimum plastic moment capacity	1.37 MNm
Expected plastic moment capacity	1.51 MNm

9.2 Lumped Masses

There are seven spans in the x-direction (Fig. 9.1). In Configuration 1, one complete span and two half spans will participate during seismic shaking. Therefore, two-seventh of the structural mass is lumped with the gantry mass. The lumped mass in Configuration 1 is $113 \times 2/7 + 60 = 92.1 \times 10^3$ kg. In Configuration 2, two half spans on either side of the column will participate during seismic shaking. Therefore, one-seventh of the structural mass is lumped with the gantry mass. The lumped mass in Configuration 2 is $113/7 + 60 = 76.2 \times 10^3$ kg.

Table 9.2   Important properties of the top flange of runway beams

Form	W36 × 160 + top plate 1.25″ × 14″
Material	ASTM A572 GR 50
Minimum yield strength, f_y	345 MPa
Expected yield strength, $1.1f_y$	379 MPa
Moment of inertia of top flange about z-axis, I_z	15,540 cm⁴
Plastic modulus of top flange of beam, Z_y	1311 cm³
Width of top flange	35.6 cm
Thickness of top flange	5.13 cm
Length of top flange, L	12.2 m
End condition	Simply supported
Lateral stiffness of top flange in configuration $1 = 12EI_z/(3La^2 - 4a^3)$	1.43 MN/m
Minimum plastic moment capacity of top flange	451 kNm
Expected plastic moment capacity of top flange	496 kNm

9.3   Ground Motion

In this chapter, the seismic response of the gantry structure is computed during the 500-year MRP ground motion at a site in the San Francisco Bay Area (SFBA). The 500-year MRP ground motion for the SFBA site was determined in Chap. 2. Figure 9.3 shows the 500-year MRP response spectra for various values of damping. The response spectra are shown in the acceleration–deformation format. These are the demand curves for the site. The peak pseudo-acceleration *PPA* is read along the vertical axis and peak deformation *PD* is read along the horizontal axis. The natural period *T* is shown by the parallel diagonal lines. The *PPA*, *PD*, and *T* are related to each other by the expression: $PD = PPA \times (T/2\pi)^2$. Only static analyses are performed in this chapter. Therefore, ground motion histories are not needed.

9.4   Elastic Analyses

First, the gantry structure is analyzed by assuming that it remains elastic (undamaged). No yielding or breaking is assumed to occur anywhere in the structure. Elastic analyses are usually linear, but in this case, even the elastic analyses are nonlinear because the stiffness of the structure changes with deformation due to the following reasons:

1. The presence of axial release in the south side wheels of the gantry causes the south side frame to engage only after the north side frame has deformed 7.62 cm.

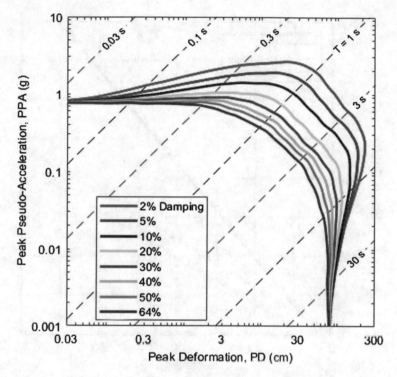

Fig. 9.3 Acceleration–deformation plots of 500-year MRP response spectra for various values of damping

2. Slender nature of the structure causes additional moment due to the weight of the gantry. This is known as the P–Δ effect.

The elastic analyses are performed for Configurations 1 and 2 by using the nonlinear-static procedure. The damping is assumed to be 2% of critical.

9.4.1   Configuration 1

The demand curve for elastic analyses is the 2% damping response spectrum in Fig. 9.3. The lumped mass for Configuration 1 is 92.1×10^3 kg. The lumped mass is slowly pushed in the north direction. This causes deflection of the beam and two columns on the north side. The stiffness of beam is 1.43 MN/m (Table 9.2) and the stiffness of two columns is $2 \times 3.85 = 7.7$ MN/m (Table 9.1). Since the beam and two columns are in series with each other in Configuration 1, the lateral stiffness of the north side frame is $1.43 \times 7.7/(1.43 + 7.7) = 1.21$ MN/m. When the deflection reaches 7.62 cm, the south side frame is also engaged. This causes the stiffness to

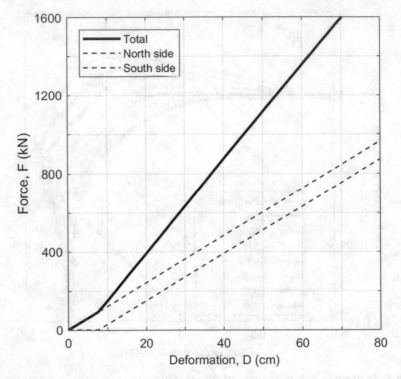

Fig. 9.4 Pushover curve for the gantry in Configuration 1 without P–Δ effect

double to 2.42 MN/m. Figure 9.4 shows the pushover curve for the structure in Configuration 1. The forces on the north and south side frames are separately shown by the two dashed lines.

The gantry structure has a significant weight and it undergoes large deflections. Therefore, the P–Δ effect should be considered for this structure. The P–Δ effect is considered as follows: (1) for each value of deformation D, force F is read from Fig. 9.4, and (2) force F is reduced by an amount $W \times D/H$, where, $W =$ weight and $H =$ height of the lumped mass from the base. Figure 9.5 shows the revised pushover curve after considering the P–Δ effect. Note a slight reduction in the stiffness.

The capacity curve is generated from the pushover curve by dividing the push-over force by the lumped mass of 92.1×10^3 kg. The capacity curve is a plot of the pseudo-acceleration versus deformation. Figure 9.6 shows the capacity curve on a linear scale. The radial tick marks indicate the "effective" period of the structure. Figure 9.7 shows the same capacity curve on a logarithmic scale. The radial tick marks in Fig. 9.6 are replaced by parallel diagonal lines in Fig. 9.7.

The demand curve is an acceleration–deformation plot of the 2% damping response spectrum. It is shown in Fig. 9.3. In Fig. 9.8, the capacity curve of Fig. 9.7 is superimposed on the 2% damping demand curve in Fig. 9.3. These two curves intersect at a deflection of 64.3 cm. This is the deflection of the gantry during

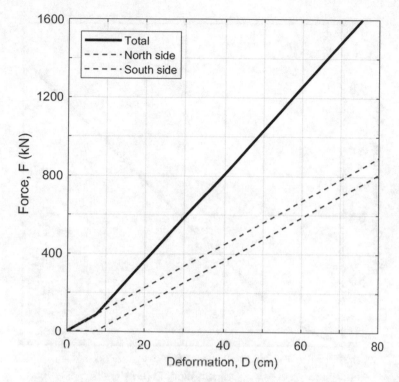

Fig. 9.5 Revised pushover curve for the gantry in Configuration 1 with P–Δ effect

the 500-year MRP ground motion. The pseudo-acceleration of the gantry is 1.49 g. The deflection of the north side frame is 64.3 cm, but the deflection of the south side frame is 64.3–7.62 = 56.7 cm.

For a deflection of 64.3 cm, the horizontal force on the north side frame is 716 kN (Fig. 9.5). This is also the force on the north side beam. The bending moment in the north side beam is 716 × 3.35/2 = 1.2 MNm. The stress in the flange of the north side beam is 1.2/15,540 × 35.6/2 × 10^6 = 1.37 GPa. This is 3.6 times the yield stress of 379 MPa for beam steel. The lateral force in the north side beam is transferred equally to the two columns at the ends of the beam. Therefore, the force in the north side column is 716/2 = 358 kN. The bending moment at the base of the north side column is 358 × 5.08/10^3 + 92,080 × 9.81/4 × 0.643 = 1.96 MNm. The stress in the north side column is 1.96 × 10^6/(84,210 × 10^{-8}) × 0.457/2/10^6 + 92,080 × 9.81/4/ 0.0267/10^6 = 540 MPa. This is 1.5 times the yield stress of 349 MPa for column steel.

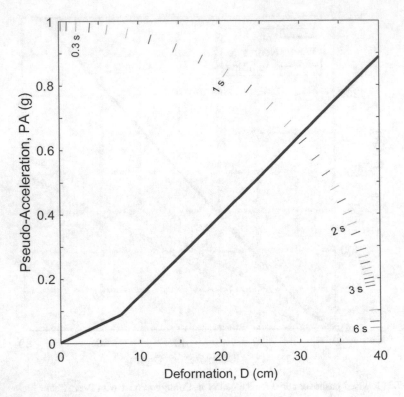

Fig. 9.6 Configuration 1 capacity curve on a linear scale

9.4.2 Configuration 2

Once again, the demand curve for elastic analysis is the 2% damping response spectrum in Fig. 9.3. The lumped mass for Configuration 2 is 76.2×10^3 kg. The lumped mass is slowly pushed in the north direction. This causes deflection of the north side column. The stiffness of the column is 3.85 MN/m (Table 9.1). When the deflection reaches 7.62 cm, the south side column is also engaged. This causes the stiffness to double to 7.7 MN/m. Figure 9.9 shows the pushover curve for the structure in Configuration 2. The forces on the north and south side columns are separately shown by the two dashed lines in Fig. 9.9. Figure 9.10 shows the revised pushover curve for Configuration 2 after considering the P–Δ effect.

The capacity curve is generated from the pushover curve by dividing the pushover force by the lumped mass of 76.2×10^3 kg. Figure 9.11 shows the capacity curve on a linear scale. The radial tick marks indicate the "effective" period of the structure. Figure 9.12 shows the same capacity curve on a logarithmic scale. The radial tick marks in Fig. 9.11 are replaced by parallel diagonal lines in Fig. 9.12.

A pushover analysis is performed in Fig. 9.13 by superimposing the capacity curve of Fig. 9.12 over the 2% damping demand curve in Fig. 9.3. These two curves

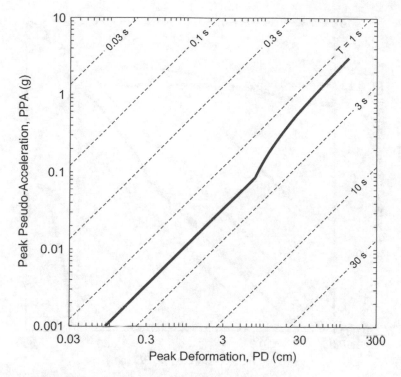

Fig. 9.7 Configuration 1 capacity curve on a logarithmic scale

intersect at a deflection of 28.9 cm. This is the deflection of the gantry during the
500-year MRP ground motion. The pseudo-acceleration of the gantry is 2.54 g. The
deflection of the north side column is 28.9 cm, but the deflection of the south side
column is $28.9 - 7.62 = 21.3$ cm. Corresponding to a deflection of 28.9 cm, the force
in the north side column is read from Fig. 9.10; it is 1.09 MN. The bending moment
at the base of the north side column is $1.09 \times 5.08 + 76.2 \times 9.81/2 \times 0.289/$
$10^3 = 5.65$ MNm. The stress in the north side column is $5.65 \times 10^6/$
$(84,210 \times 10^{-8}) \times 0.457/2/10^9 + 76.2 \times 9.81/2/0.0267/10^6 = 1.55$ GPa. This is
4.4 times the yield stress of 349 MPa for column steel.

9.4.3   Conclusions from Elastic Analyses

The results of the elastic analyses show that yielding will occur in the structure
during the 500-year MRP ground motion. To determine the performance of the
structure during the 500-year MRP ground motion, the extent of yielding needs to be
calculated. The extent of yielding in frame elements (beams and columns) is
measured by the plastic hinge rotations. An inelastic analysis is needed to estimate
the plastic hinge rotations in beam and column elements.

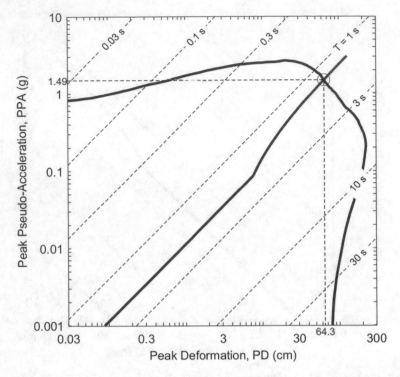

Fig. 9.8 Elastic pushover analysis of the gantry in Configuration 1

9.5 Inelastic Analyses

The next set of analyses are inelastic. In these analyses, plastic hinges are allowed to form in the structure and the rotations of plastic hinges are monitored. Plastic rotations affect the stiffness of the structure and they also affect its damping because energy is dissipated in the plastic hinges. Once again, the analyses are performed by the nonlinear-static procedure.

9.5.1 Configuration 1

Figure 9.14 shows the inelastic pushover curve for the structure in Configuration 1. As the lumped mass is slowly pushed in the north direction, the entire load is first resisted by the north side frame. This causes deflection of the north side beam and two columns at the ends of beam. When the deflection reaches 7.61 cm, the south side frame is also engaged and the stiffness of the system doubles. With additional deflection, the bending moment in the north side beam continues to rise until it reaches its plastic moment capacity. At that point, two plastic hinges appear in the

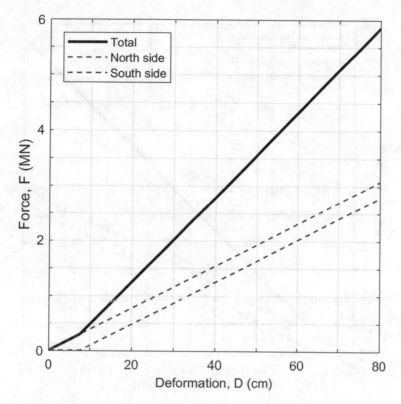

Fig. 9.9 Pushover curve for the gantry in Configuration 2 without P–Δ

north side beam at the wheel locations. This causes the stiffness of the system to reduce. At a later point, plastic hinges also appear in the south side beam and the stiffness drops once again. In this analysis, strain hardening is ignored. Negative slope (stiffness) of the pushover curve after the formation of plastic hinges in the south side beam is due to P–Δ effect. The gantry collapses at 79.7 cm deformation when the plastic hinge rotation in the north beam reaches its limiting value of 0.179 radian (Section 9.5.2).

The capacity curve is obtained by dividing the pushover force by the lumped mass of 92.1×10^3 kg. A linear plot of the capacity curve is shown in Fig. 9.15. The radial tick marks in Fig. 9.15 indicate the "effective" period of the structure. Figure 9.16 shows the same capacity curve on a logarithmic scale. The radial tick marks in Fig. 9.15 are replaced by diagonal parallel lines in Fig. 9.16.

In Chap. 3, the seismic toughness ST of a structure was defined as the area under the linear plot of the capacity curve (Fig. 9.15). The seismic toughness of the gantry in Configuration 1 is $ST=2.79$ (m/s)^2. This is 10 to 14% higher than the value for ductile moment frames discussed in Chapters 3 and 4. Therefore, the seismic toughness of the gantry in Configuration 1 is quite high.

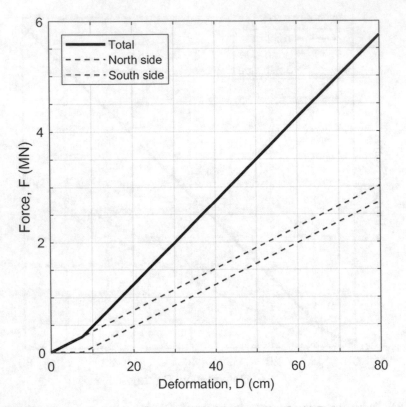

Fig. 9.10 Revised pushover curve for the gantry in Configuration 2 with P–Δ

A certain amount of energy is dissipated due to yielding. Figure 9.17 shows cyclic force–deformation relationship for the structure undergoing a peak deformation of 50 cm. The area of the shaded loop on the left side is the energy dissipated in each cycle. It is denoted as E_h. The area of the shaded triangle on the right side is the "strain energy." It is denoted as E_s. The hysteretic damping of the system is obtained from the relation [4]:

$$\zeta_h = \frac{E_h}{4\pi E_s}$$

The hysteretic damping ζ_h can be calculated for cycles of various amplitudes. Figure 9.18 shows a plot of hysteretic damping versus deformation. This is the "raw" damping curve which needs some adjustments to account for the fact that a structure experiences cycles of many different amplitudes during the ground shaking. Since the damping for smaller amplitude cycles is less, the damping is adjusted as follows. For $D = 40$ cm, the damping is 0.3 (30% of critical) according to the "raw" damping curve (Fig. 9.18). The average (adjusted) damping for deformations between 0 and 40 cm is calculated by taking the area under the damping curve up to 40 cm and

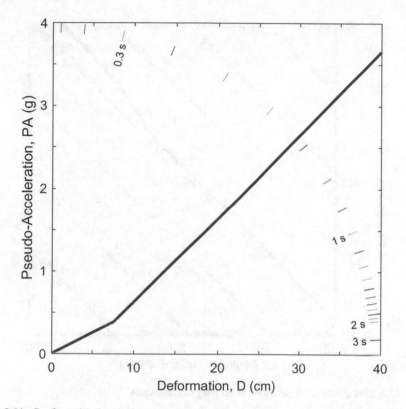

Fig. 9.11 Configuration 2 capacity curve on a linear scale

dividing that area by 40 cm (Fig. 9.19). This gives the adjusted damping of 3.48/ 40 = 0.087 (8.7% of critical). The adjusted damping is similarly computed for other values of D. Finally, the damping is not allowed to drop below 2% of critical to account for other sources of energy dissipation besides plastic yielding. Figure 9.20 shows the adjusted damping curve for the system; this will be used to complete the nonlinear-static analysis.

The damping of the system is not known prior to the analysis. In Fig. 9.21, the capacity curve (Fig. 9.16) is superimposed on demand curves for 5, 10, 20, 30, and 50% of critical damping (Fig. 9.3). The intersections of capacity curve with demand curves provide deformations for various assumed values of damping. Note that the capacity curve stops short of the 5% damping demand curve. This implies that the gantry structure would collapse under the 500-year MRP ground motion if the damping were only 5% of critical. Figure 9.22 shows the deformations for various assumed values of damping. This is the deformation-versus-damping curve even though the deformation is shown along the horizontal axis. Higher the damping, smaller is the deformation.

In Fig. 9.23, the deformation-versus-damping curve of Fig. 9.22 is superimposed on the damping-versus-deformation curve of Fig. 9.20. These two curves intersect at

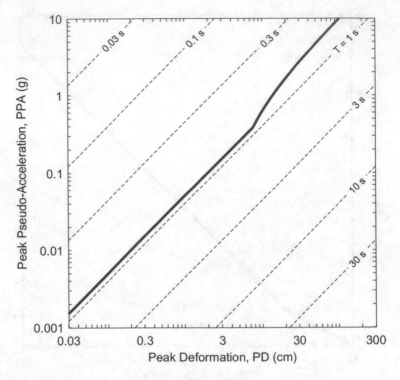

Fig. 9.12 Configuration 2 capacity curve on a logarithmic scale

a deformation of 49 cm. This is the deformation of the north side frame during the 500-year MRP ground motion. The elastic deformation of the north side frame is 19.7 cm (Fig. 9.14). Therefore, inelastic deformation of the north side frame is 49–19.7 = 29 cm. The inelastic deformation occurs only in the beam at the wheel locations. The total plastic rotation in the north side beam is 29/335 = 0.087 radian.

9.5.2 Performance of Gantry in Configuration 1

According to Table 9.7.1 of ASCE 41 [3], the acceptable plastic rotation in a compact beam cross section is $2.25\theta_y$ for immediate occupancy (IO) and $11\theta_y$ for collapse prevention (CP) limit state. The chord rotation at yield, as per Fig. 9.3(a) in ASCE 41 [3], is:

$$\theta_y = \frac{M_p \cdot l}{3EI} \tag{9.1}$$

Substituting M_p = 451 kNm (Table 9.2); $l = a$= 3.35 m (Fig. 9.2), E = 200 GPa, $I = I_z$=15,540 cm^4 (Table 9.2), gives θ_y = 451 × 3.35/(3 × 200 × 10^6 × 15,540 ×

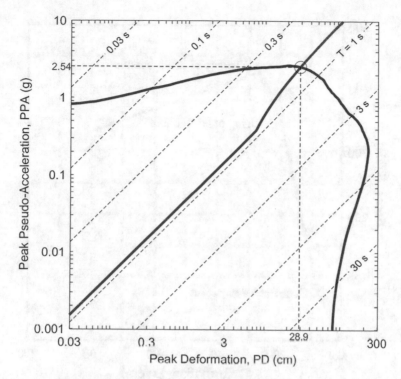

Fig. 9.13 Elastic pushover analysis of the gantry in Configuration 2

10^{-8}) = 0.0163 radian. Therefore, the acceptable plastic rotation is $2.25 \times 0.0163 =$ 0.0367 radian for IO and $11 \times 0.0163 = 0.179$ radian for CP. The calculated plastic rotation of 0.087 radian is greater than the acceptable plastic rotation for IO but less than the acceptable plastic rotation for CP. Therefore, the beam will suffer extensive damage but not break during the 500-year MRP ground motion. The gantry will not remain operational after the 500-year MRP ground motion at the site. The plastic rotation can be reduced to the acceptable level by increasing the size of the top flange.

9.5.3 Configuration 2

Figure 9.24 shows the pushover curve for the system in Configuration 2. As the lumped mass is slowly pushed in the north direction, the entire load is first transmitted to the north side column. When the deflection reaches 6.7 cm, a plastic hinge is formed at the base of north side column. At a deflection of 7.62 cm, the south side column is also engaged. At a deflection of 14.2 cm, a plastic hinge appears at the base of south side column as well. Negative slope (stiffness) of the pushover curve after the formation of hinge in the south side column is due to P–Δ effect.

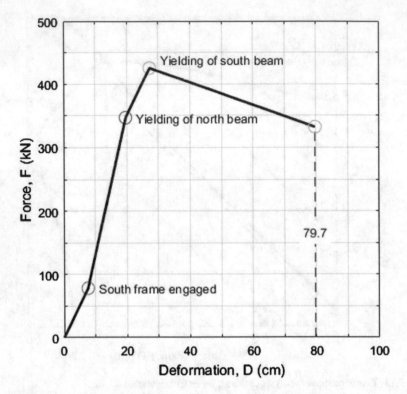

Fig. 9.14 Inelastic pushover curve for the gantry structure in Configuration 1

The capacity curve is obtained by dividing the pushover force by the lumped mass of 76.2×10^3 kg for Configuration 2. A linear plot of the capacity curve is shown in Fig. 9.25. The radial tick marks in Fig. 9.25 indicate the "effective" period of the structure. Figure 9.26 shows a logarithmic plot of the same capacity curve. The radial tick marks in Fig. 9.25 are replaced by diagonal parallel lines in Fig. 9.26.

The seismic toughness ST of a structure is the area under the linear plot of the capacity curve (Fig. 9.25). The seismic toughness of the gantry in Configuration 2 is $ST=7.86$ (m/s)^2. This is 2.8 times the ST in Configuration 1 and 3.1 to 3.2 times the ST for ductile moment frames discussed in Chapters 3 and 4. Therefore, the seismic toughness of the gantry in Configuration 2 is very high. This is due to high allowable plastic rotation in the compact cross section of the column.

Figure 9.27 shows cyclic force–deformation relationship for the structure undergoing a peak deformation of 50 cm. The area of the shaded loop on the left side is the energy dissipated in each cycle. It is denoted by E_h. The area of the shaded triangle on the right side is the "strain energy." The damping curve for the system is generated by using the same procedure as for Configuration 1. Figure 9.28 shows the adjusted damping curve for nonlinear-static analysis in Configuration 2.

In Fig. 9.29, the capacity curve (Fig. 9.26) is superimposed on demand curves for various values of damping (Fig. 9.3). The intersections of capacity curve with

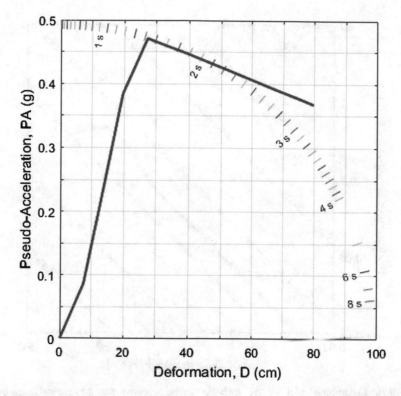

Fig. 9.15 Linear plot of inelastic capacity curve for the gantry structure in Configuration 1

demand curves provide deformations for various assumed values of damping. Figure 9.30 shows the deformations for various assumed values of damping. This is the deformation-versus-damping curve for Configuration 2. Higher the damping, smaller is the deformation. To determine the equilibrium condition, the deformation-versus-damping curve of Fig. 9.30 is superimposed on the damping-versus-deformation curve of Fig. 9.28. These two curves intersect at a deformation of 24 cm (Fig. 9.31). This is the deformation of the north side column during the 500-year MRP ground motion. The elastic deformation of the north side column is 6.7 cm (Fig. 9.24). Therefore, inelastic deformation of the north side column is 24–6.7 = 17 cm. The plastic rotation in the north side column is 17/508 = 0.034 radian (Fig. 9.32).

9.5.4 Performance of Gantry in Configuration 2

According to Table 9.7.1 of ASCE 41 [3], the acceptable plastic rotation for the column is 0.0589 radian for immediate occupancy (IO) and 0.3 radian for collapse prevention (CP).

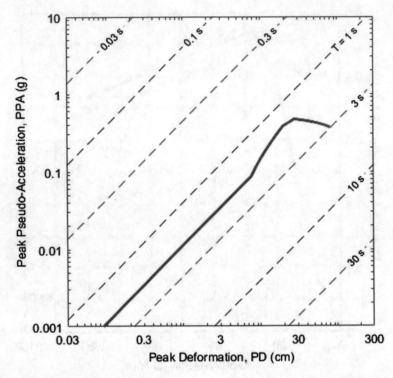

Fig. 9.16 Logarithmic plot of the inelastic capacity curve for the gantry structure in Configuration 1

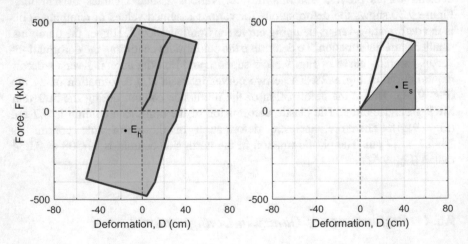

Fig. 9.17 Hysteretic and strain energies for the system in Configuration 1 for a 50-cm deformation cycle

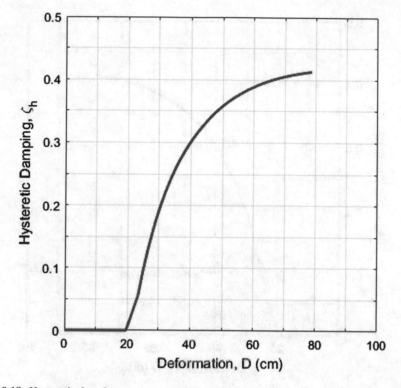

Fig. 9.18 Hysteretic damping curve

The calculated plastic rotation of 0.034 radian is less than the acceptable value for IO but the gantry cannot be expected to remain operational if any yielding occurs in the columns. Therefore, the gantry will not be collapsed by the 500-year MRP ground motion, but it will not remain operational.

9.6   Design of Column Connection

The connection at the base of column should be strong enough to allow full plastic rotation to occur in the column. At its base, each column is provided with 61 cm high vertical stiffeners. To calculate the moment demand on the connection, the plastic hinge can be assumed to form at 61 cm height, above the stiffeners. The moment in the connection, at the base of column, will be greater than the plastic moment capacity of the column.

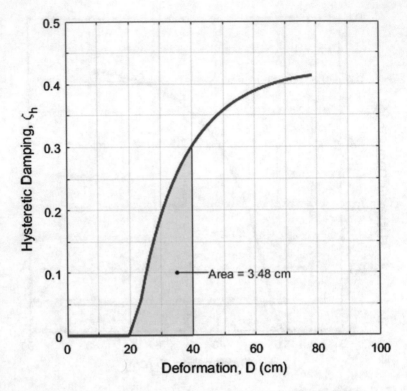

Fig. 9.19 Adjustment of damping curve

9.7   Summary

1. For unique structures such as a gantry, prescriptive approach cannot be justified because there aren't enough experience data to justify an empirical approach.
2. A pure analysis based on engineering mechanics principles is the best option.
3. To ensure ductile performance, all connections (including the base anchorage) should be stronger than the connected members.
4. Plastic rotations in ductile elements provide a measure of damage.
5. Plastic rotations can be reduced by increasing the strength of ductile elements, but brittle elements should always be stronger than ductile elements.
6. Plastic rotations should be calculated by using lower bound estimates of yield strength. The moments in connections and foundations should be calculated by using upper bound estimates of yield strength.

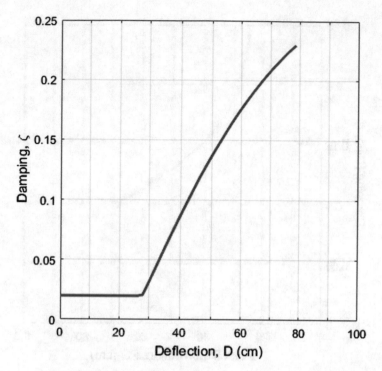

Fig. 9.20 Adjusted damping curve for Configuration 1

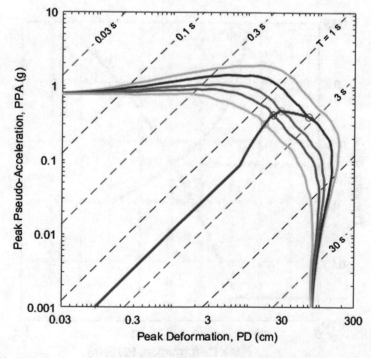

Fig. 9.21 Capacity curve superimposed on demand curves for 5, 10, 20, 30, and 50% of critical damping

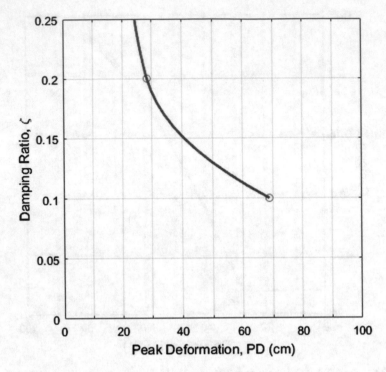

Fig. 9.22 Deformations for various assumed values of damping

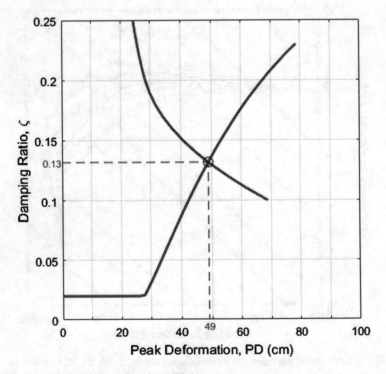

Fig. 9.23 Damping and deformation at equilibrium

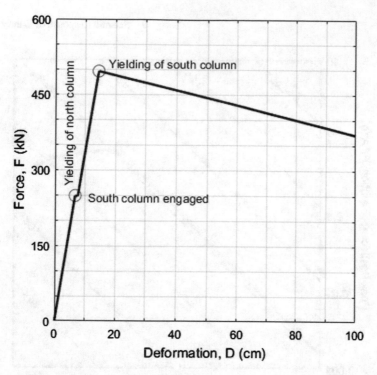

Fig. 9.24 Inelastic pushover curve for Configuration 2

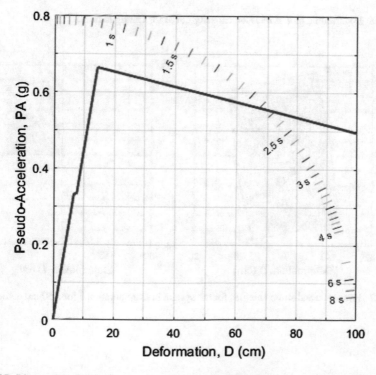

Fig. 9.25 Linear plot of inelastic capacity curve for Configuration 2

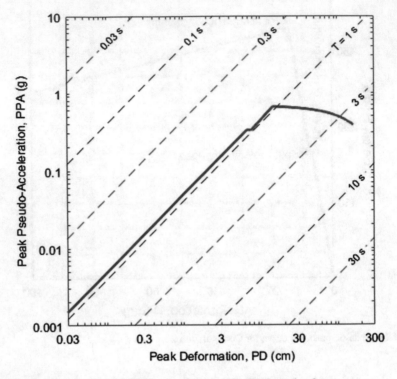

Fig. 9.26 Logarithmic plot of inelastic capacity curve for Configuration 2

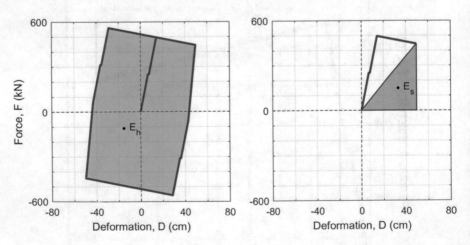

Fig. 9.27 Hysteretic and strain energies for the system in Configuration 2 for a 50-cm deformation cycle

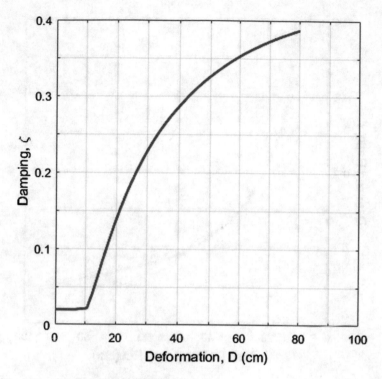

Fig. 9.28 Adjusted damping curve for Configuration 2

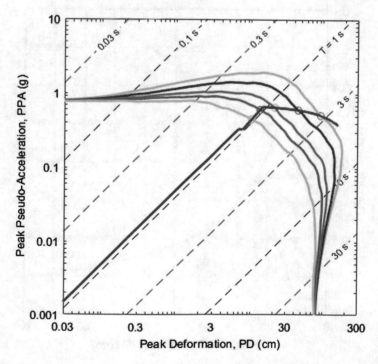

Fig. 9.29 Capacity curve for Configuration 2 superimposed on demand curves for 5, 10, 20, 30, and 50% of critical damping

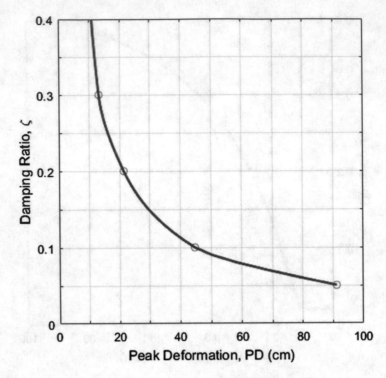

Fig. 9.30 Deformations for various values of damping in Configuration 2

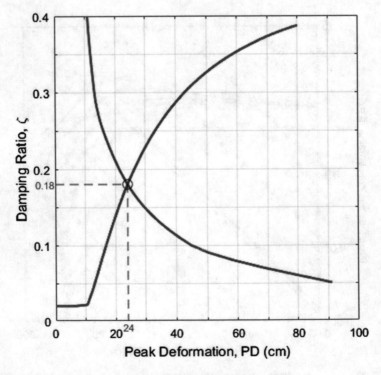

Fig. 9.31 Deformation and damping at equilibrium in Configuration 2

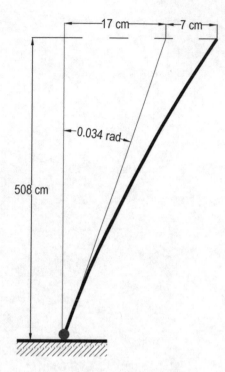

Fig. 9.32 Plastic rotation at the base of column in Configuration 2

References

1. AISC (2016). *"Specification for Structural Steel Buildings."* ANSI/AISC 360-16.
2. MathWorks. (2020). *MATLAB Version 9.8.0.1417392 (R2020a)*. Natick, MA: The MathWorks, Inc.
3. ASCE. (2017). Seismic evaluation and retrofit of existing buildings. In *ASCE 41–17*. Reston, VA: American Society of Civil Engineers.
4. Chopra, A. K. (2017). *Dynamics of structures*. Fifth Edition: Pearson Publication.

Index

© Springer Nature Switzerland AG 2021
P. K. Malhotra, *Seismic Analysis of Structures and Equipment*,
https://doi.org/10.1007/978-3-030-57858-9

Printed in the United States
by Baker & Taylor Publisher Services